Quantum Enigma

Quantum Enigma

Physics Encounters Consciousness

Second Edition

Bruce Rosenblum *and* Fred Kuttner

OXFORD
UNIVERSITY PRESS

OXFORD
UNIVERSITY PRESS

Oxford University Press, Inc., publishes works that further
Oxford University's objective of excellence
in research, scholarship, and education.

Oxford   New York
Auckland   Cape Town   Dar es Salaam   Hong Kong   Karachi
Kuala Lumpur   Madrid   Melbourne   Mexico City   Nairobi
New Delhi   Shanghai   Taipei   Toronto

With offices in
Argentina   Austria   Brazil   Chile   Czech Republic   France   Greece
Guatemala   Hungary   Italy   Japan   Poland   Portugal   Singapore
South Korea   Switzerland   Thailand   Turkey   Ukraine   Vietnam

Copyright © 2011 by Bruce Rosenblum and Fred Kuttner

Published by Oxford University Press, Inc.
198 Madison Avenue, New York, New York 10016

www.oup.com

Oxford is a registered trademark of Oxford University Press

All rights reserved. No part of this publication may be reproduced,
stored in a retrieval system, or transmitted, in any form or by any means,
electronic, mechanical, photocopying, recording, or otherwise,
without the prior permission of Oxford University Press, Inc.

Library of Congress Cataloging-in-Publication Data
Rosenblum, Bruce.
Quantum enigma : physics encounters consciousness / Bruce Rosenblum and
Fred Kuttner. — 2nd ed.
    p. cm.
ISBN 978-0-19-975381-9 (pbk. : alk. paper)
1. Quantum theory.   2. Science—Philosophy.
I. Kuttner, Fred.   II. Title.
QC174.13.R67 2011
530.12—dc22   2010019465

5 7 9 8 6 4
Printed in the United States of America
on acid-free paper

We dedicate our book to the memory of John Bell, perhaps the leading quantum theorist of the latter half of the twentieth century. His writings, lectures, and personal conversations have inspired us.

Is it not good to know what follows from what, even if it is not necessary FAPP? [FAPP is Bell's suggested abbreviation of "for all practical purposes."] Suppose for example that quantum mechanics were found to resist precise formulation. Suppose that when formulation beyond FAPP is attempted, we find an unmovable finger obstinately pointing outside the subject, to the mind of the observer, to the Hindu scriptures, to God, or even only Gravitation? Would that not be very, very interesting?

—John Bell

Acknowledgments

During the preparation of our book we have greatly benefited from the suggestions, criticism, and corrections offered us by those who have read chapters as they were being prepared and revised. We gratefully acknowledge the help of Leonard Anderson, Phyllis Arozena, Donald Coyne, Reay Dick, Carlos Figueroa, Freda Hedges, Nick Herbert, Alex Moraru, Andrew Neher, and Topsy Smalley.

We warmly thank our former editor, Michael Penn, for his insightful advice and support, and our present editor, Phyllis Cohen, for her insightful advice, continued support, and encouragement of further projects. We are grateful to our former production editor, Stephanie Attia, for her valuable suggestions. And thanks to production editor Amy Whitmer for her efficient help with the second edition.

All along, our agent, Faith Hamlin, has given us crucial advice and warm encouragement. We very much appreciate her involvement in our book.

Contents

Preface to the Second Edition

Quantum mechanics is stunningly successful. Not a single prediction of the theory has ever been wrong. One-third of our economy depends on products based on it. However, quantum mechanics also displays an enigma. It tells us that physical reality is created by observation, and it has "spooky actions" instantaneously influencing events far from each other–without any physical force involved. Seen from a human perspective, quantum mechanics has physics encountering consciousness.

Our book describes the completely undisputed experimental facts and the accepted explanation of them by the quantum theory. We discuss today's contending interpretations, and how each encounters consciousness. Fortunately, the quantum enigma can be deeply explored in non-technical language. The mystery presented by quantum mechanics, which physicists call the "quantum measurement problem," appears right up front in the simplest quantum experiment.

In recent years, investigations into the foundations, and the mysteries, of quantum mechanics have surged. Quantum phenomena are ever more apparent in fields ranging from computer engineering, to biology, to cosmology. This second edition includes recent advances in both understanding and applications. Our use of the book in large classes and small seminars has enabled us to improve our presentation. Improvement has also benefited from the response of readers, other instructors who have used the book, and the comments of reviewers. We intend to expand and update coverage of certain topics on our book's website: quantumenigma.com.

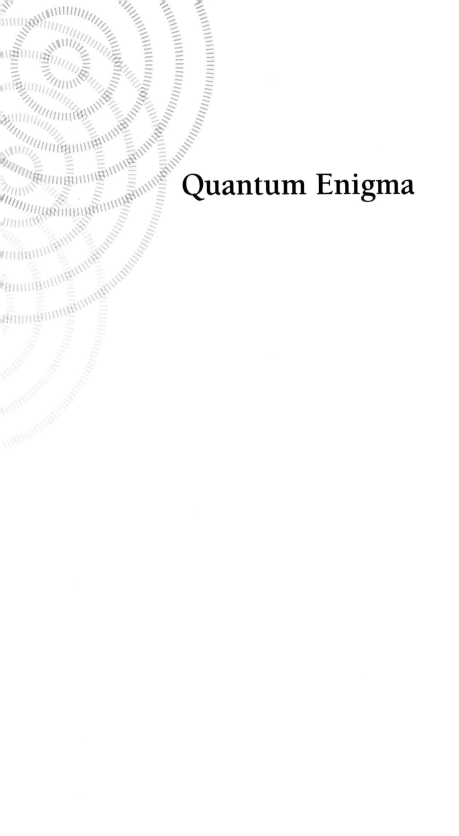

Quantum Enigma

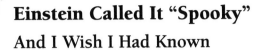

Einstein Called It "Spooky"
And I Wish I Had Known

I have thought a hundred times as much about the quantum problem as I have about general relativity theory.

—Albert Einstein

I cannot seriously believe in [quantum theory] because
. . . physics should represent a reality in time and space, free from spooky action at a distance.

—Albert Einstein

I was visiting friends in Princeton one Saturday in the 1950s when our host asked his son-in-law, Bill Bennett, and me (Bruce) if we'd like to spend the evening with his friend, Albert Einstein. Two awed physics graduate students soon waited in Einstein's living room as he came downstairs in slippers and sweatshirt. I remember tea and cookies but not how the conversation started.

Einstein soon asked about our quantum mechanics course. He approved of our professor's choice of David Bohm's book as the text, and he asked how we liked Bohm's treatment of the strangeness quantum theory implied. We couldn't answer. We'd been told to skip that part of the book and concentrate on the section titled "The Mathematical Formulation of the Theory." Einstein persisted in exploring our thoughts about what the theory really meant. But the issues that concerned him were unfamiliar to us. Our quantum physics courses focused on the *use* of the theory,

3

not its meaning. Our response to his probing disappointed Einstein, and that part of our conversation ended.

It would be many years before I understood Einstein's concern with the mysterious implications of quantum theory. I did not know that back in 1935 he had startled the developers of quantum theory by pointing out that the theory required an observation at one place to *instantaneously* influence what happened far away *without involving any physical force*. He derided this as "spooky action" that could not actually exist.

Einstein was also bothered by the theory's claim that if you observed a small object, an atom, say, to be someplace, it was your *looking* that caused it to be there. Does that apply to big things? In principle, yes. Ridiculing quantum theory, Einstein once asked a fellow physicist, only *half*-jokingly, if he believed the moon was there only when he looked at it. According to Einstein, if you took quantum theory seriously, you denied the existence of a physically real world independent of its observation. This is a serious charge. Quantum theory is not just one of many theories in physics. It is the framework upon which *all* of physics is ultimately based.

Our book focuses on the mysterious implications of quantum theory that bothered Einstein, from his initial proposal of the quantum in 1905 to his death a half-century later. But for many years after that evening with Einstein, I hardly thought about the quantum weirdness, which physicists call "the measurement problem." As a graduate student, I puzzled about the related "wave–particle duality." It's the paradox that, looking one way, you could demonstrate an atom to be a compact object concentrated in one place. However, looking differently, you could demonstrate exactly the *opposite*. You could show that the atom was *not* a compact object, that it was a wave spread out over a wide region. That contradiction puzzled me, but I assumed that if I spent some hours thinking it through, I'd see it all clearly–the way my professors seemed to. As a graduate student, I had more pressing things to do. My Ph.D. thesis involved lots of quantum theory, but like most physicists, I had little concern with the theory's deeper meaning, which I then didn't realize goes well beyond mere "wave–particle duality."

After a decade in industrial physics research and research management, I joined the faculty at the University of California, Santa Cruz (UCSC). Teaching a physics course for liberal arts students, the mysteries of quantum mechanics intrigued me. A weeklong conference in Italy on the foundations of quantum mechanics left me hooked on what I was unprepared to talk about that long-ago evening in Princeton.

When I (Fred) encountered quantum mechanics in my junior year at MIT, I wrote Schrödinger's equation across the page of my notebook, excited to see the equation that governed everything in the universe. Later I puzzled about the quantum assertion that an atom's north pole could point in more than one direction at the same time. I wrestled with this for a while but gave up, figuring I'd understand it after I learned more.

For my Ph.D. dissertation I did a quantum analysis of magnetic systems. I had become facile in *using* quantum theory, but I had no time to think about what it *meant*. I was too busy trying to publish papers and get my degree. After working with a couple of hi-tech companies, I joined the physics faculty at UCSC.

When the two of us started to explore the boundary where physics meets speculative philosophy, our physics colleagues were surprised. Our previous research areas were quite conventional, even practical. (There is more about our backgrounds in industrial and academic research, and contact information, on our book's website: www.quantumenigma.com.)

The Skeleton in Physics' Closet

Quantum theory is stunningly successful. Not a single one of the theory's predictions has ever been shown wrong. One-third of our economy depends on products based on it. However, the worldview demanded by quantum theory is not only stranger than we might suppose, it's stranger than we *can* suppose. Let's see why.

Most of us share these commonsense intuitions: A single object can't be in two far-apart places at once. And, surely, what someone decides to do here cannot instantly affect what happens someplace far away. And doesn't it go without saying that there's a real world "out there," whether or not we look at it? Quantum mechanics challenges each of these intuitions. J. M. Jauch tells us: "For many thoughtful physicists, [the deeper meaning of quantum mechanics] has remained a kind of skeleton in the closet."

We started out telling of Einstein's troubled concern with quantum theory. What *is* quantum theory? Quantum theory was developed early in the twentieth century to explain the *mechanics*, the mechanism, governing the behavior of atoms. Early on, it was discovered that the energy of an object could change only by a discrete quantity, a *quantum*, hence "quantum mechanics." "Quantum mechanics" includes both the experimental observations and the quantum *theory* explaining them.

Quantum theory is at the base of every natural science from chemistry to cosmology. We need quantum theory to understand why the sun shines, how TV sets produce pictures, why grass is green, and how the universe developed from the Big Bang. Modern technology is based on devices designed with quantum theory.

Prequantum physics, "*classical* mechanics," or "classical physics," also sometimes called "Newtonian physics," is usually an excellent approximation for objects much larger than molecules, and it's typically much simpler to use than quantum theory. It is, however, only an approximation. It does not work at all for the atoms that everything is made of. Nevertheless, classical physics is basic to our conventional wisdom, our Newtonian worldview. But we now know this classical worldview is fundamentally flawed.

Since ancient times, philosophers have come up with esoteric speculations on the nature of physical reality. But before quantum mechanics, one had the logical option of rejecting such theorizing and holding to a straightforward, commonsense worldview. Today, quantum experiments deny a commonsense physical reality. It is no longer a logical option.

Might a worldview suggested by quantum mechanics have relevance beyond science? Consider earlier discoveries that did have such relevance: Copernicus's realization that Earth was not the center of the cosmos, or Darwin's theory of evolution. The relevance of quantum mechanics is, in a sense, more immediate than Copernican or Darwinian ideas, which deal with the far away or long ago. Quantum theory is about the here and now. It even encounters the essence of our humanity, our consciousness.

Why, then, hasn't quantum mechanics had the intellectual and societal impact of those earlier insights? Perhaps because they are easier to comprehend. They are certainly *much* easier to believe. You can roughly summarize the implications of Copernicus or Darwin in a few sentences.

To the modern mind, at least, they seem reasonable. Try summarizing the implications of quantum theory, and what you get sounds mystical.

We risk a rough summary anyway. Quantum theory tells that the observation of an object can instantaneously influence the behavior of another greatly distant object—*even if no physical force connects the two.* These are the influences Einstein rejected as "spooky actions," but they have now been demonstrated to exist. Quantum theory also tells us that an object can be in two places at the same time. Its existence at the particular place where it happens to be found becomes an actuality *only upon its observation.* Quantum theory thus denies the existence of a physically real world independent of its observation. (We'll see "observation" to be a tricky and controversial concept.)

Strange quantum phenomena can be *directly* demonstrated only for small objects. Classical physics describes the reasonable behavior of big things to an *extremely* good approximation. But the big things are made up of the small things. As a worldview, classical physics just does not work.

Classical physics explains the world quite well; it's just the "details" it can't handle. Quantum physics handles the "details" perfectly; it's just the world it can't explain. You can see why Einstein was troubled.

Erwin Schrödinger, a founder of modern quantum theory, told his famous cat story to emphasize that quantum theory says something "absurd." Schrödinger's unobserved cat, according to quantum theory, was simultaneously dead **and** alive until your observation of it *causes* it to be either dead or alive. Here's something even harder to accept: Finding the cat dead creates the history of its developing rigor mortis. Finding it alive creates the history of its developing hunger. *Backward in time.*

The enigma posed by quantum theory has challenged physicists for eight decades. Perhaps the particular expertise and talents of physicists does not *uniquely* qualify us for its comprehension. We physicists might therefore approach the problem with modesty, though we find that hard.

Remarkably, the quantum enigma can be presented essentially full-blown without involving much physics background. Might someone unencumbered by years of training in the *use* of quantum theory have a new insight? It was a child who pointed out that the emperor wore no clothes.

Controversy

Our book originated with a wide-ranging physics course for liberal arts students that in its last weeks focused on the mysteries of quantum mechanics. When I (Bruce) first proposed the course at a department meeting, that final focus prompted a faculty member to object:

> Though what you are saying is correct, presenting this material to nonscientists is the intellectual equivalent of allowing children to play with loaded guns.

That objecting faculty member, a good friend, had a valid concern: Some people, seeing the solid science of physics linked with the mystery of the conscious mind, might become susceptible to all sorts of pseudo-scientific nonsense. My response was that we'd teach "gun-safety": We'd emphasize the scientific method. The course was approved. Fred now teaches it, and it's become the most popular course in our department.

Let's note straightaway that the encounter with consciousness in our title does not imply "mind control," that your thoughts alone can *directly* control the physical world. Do the undisputed results of the quantum experiments we describe imply a mysterious role for consciousness in the *physical* world? It's a hotly debated question arising at a boundary of the physics discipline.

Since our book focuses on that boundary, where the quantum enigma emerges, it is necessarily a controversial book. However, absolutely *nothing* we say about quantum mechanics itself is controversial. It is the mystery these results imply *beyond* the physics that is controversial. For many physicists, this baffling weirdness is best not talked about. Physicists (including ourselves) can be uncomfortable with their discipline encountering something as "unphysical" as consciousness. Though the quantum facts are not in dispute, the meaning *behind* those facts, what quantum mechanics tells us about our world, is hotly debated. Addressing them in a physics department, especially in a physics class or to a non-technical audience, will incur the disapproval of some faculty. (Physicists, of course, are not alone in their discomfort with the issue of consciousness arising mysteriously in the discussion of physical phenomena. It can challenge the worldview of any of us.)

An Einstein biographer tells that back in the 1950s a non-tenured faculty member in a physics department would endanger a career by showing any interest in the strange implications of quantum theory. Times are changing. Exploration of the fundamental issues in quantum mechanics, which cannot avoid encountering consciousness, increases today and extends beyond physics to psychology, philosophy, and even computer engineering.

Since quantum theory works perfectly for all *practical* purposes, some physicists deny there's any problem. Such denial abandons to the purveyors of pseudo-science the aspects of quantum mechanics that understandably most intrigue non-physicists. The movie *What the Bleep?* is an example of the pseudo-science we deplore. (If you're unfamiliar with *Bleep*, see our comment early in chapter 15.) The *real* quantum enigma is more bizarre and more profound than the "philosophies" such treatments espouse. Understanding the real quantum mystery requires a bit more mental effort, but it's worth it.

At a physics conference attended by several hundred physicists (including the two of us), an argument broke out in the discussion period after a talk. (The heated across-the-auditorium debate was reported in the *New York Times* in December 2005.) One participant argued that because of its weirdness, quantum theory had a problem. Another vigorously denied there was a problem, accusing the first of having "missed the point." A third broke in to say, "We're just too young. We should wait until 2200 when quantum mechanics is taught in kindergarten." A fourth summarized the argument by saying, "The world is not as real as we think." Three of these arguers have Nobel Prizes in Physics, and the fourth is a good candidate for one.

This argument recalls an analogy that reflects our own bias. A couple is in marriage counseling. The wife says, "There's a problem in our marriage." Her husband disagrees, saying, "There's *no* problem in our marriage." The marriage counselor knows who's right.

Interpreting Quantum Theory

In the last twenty years of his life, Einstein's continued challenging of quantum theory was often dismissed as his being out of touch with modern physics. He was indeed wrong in denying the reality of the "spooky action" he discovered to lurk in quantum theory. Its existence, now called "entanglement," has been demonstrated. Nevertheless, Einstein is today recognized as the theory's most prescient critic. His constant claim that the theory's weirdness must not be brushed aside is borne out by today's proliferation of wild interpretations of quantum theory.

In chapter 15 we describe several contending views, interpretations, of what quantum mechanics is telling us about the physical world—and, perhaps, about us. These are all serious proposals developed with extensive mathematical analysis. They variously suggest observation creating a physical reality, the existence of many parallel worlds with each of us in each of them, a universal connectedness, the future affecting the past, a reality beyond physical reality, and even a challenge to free will.

At the boundary where physics no longer compels consensus, the meaning of quantum theory is controversial. Most interpretations of what's going on show how the issue of consciousness can be ignored for all *practical* purposes. However, in exploring the theory's foundations, most contemporary experts admit a mystery, usually one encountering consciousness. Although it is our most intimate experience, consciousness is ill defined. It's something physics can't treat, but can't ignore.

Physics Nobel Laureate Frank Wilczek recently commented:

> The relevant literature [on the meaning of quantum theory] is famously contentious and obscure. I believe it will remain so until someone constructs, within the formalism of quantum mechanics, an "observer," that is, a model entity whose states correspond to a recognizable caricature of conscious awareness.
> . . . That is a formidable project, extending well beyond what is conventionally considered physics.

As we present the undisputed facts, and emphasize the enigma they challenge us with, we propose no resolution of the enigma. We rather offer readers a basis for their own pondering. Remarkably, this controversial issue can be understood with little prior knowledge of physics.

2

The Visit to Neg Ahne Poc
A Quantum Parable

If you're going to ham it up, go the whole hog.

—G. I. Gurdjieff

A few chapters will go by before we encounter the enigma posed by quantum mechanics. But let's start out with a look at the paradox. With today's technology we can display the quantum enigma only with tiny objects. But quantum mechanics supposedly applies to everything.

So we begin by telling a story in which a physicist visits Neg Ahne Poc, a land with a magical technology that allows displaying something *like* the quantum enigma with large objects, a man and a woman, instead of atoms. Our parable tells of something impossible in the real world, but watch for what baffles our visitor to Neg Ahne Poc. His *bafflement* is the point of our parable. In later chapters you should experience a similar bafflement.

Prologue by Our Self-Assured Visitor to
Neg Ahne Poc

Let me tell you why I'm slogging up this steep trail. Since quantum mechanics can make Nature appear mystical, some people can be misled into accepting supernatural foolishness.

Last month I was with some usually sensible friends in California. People there, however, seem particularly susceptible to quantum nonsense. My friends spoke of the "Rhob" in Neg Ahne Poc, a village high in the Hima-Ural Mountains. They claimed this shaman could display quantum-like phenomena with large objects. That's ridiculous, of course!

When I explained to them that such a demonstration is impossible, they accused me of being a closed-minded scientist. I was challenged to investigate. One of them, a dot-com billionaire, who admits that selling his company only months before the bust was just dumb luck, offered to fund my trip. Colleagues in the physics department urged me not to waste my time on a wild-goose chase, that I had better do serious physics and publish if I'm going to get tenure. But I believe that a public-spirited scientist should expend some effort investigating unjustified notions to prevent their propagation. So here I am.

I'll look into this stuff with a completely open mind. I'll then debunk this nonsense when I get home. But while I'm in Neg Ahne Poc, I'll be discreet. This shaman's trickery is likely part of the local religion.

The trail becomes less steep and broadens to end suddenly in a modest plaza. Our visitor has arrived in Neg Ahne Poc. He is relieved to see that his friends' long-distance arrangements have worked. His arrival is expected. He is warmly greeted by the Rhob and a small group of villagers.

� Greetings, Curious Questioner, Careful Experimenter. You are a welcome visitor to our village.

☦ Thank you, thank you very much. I appreciate the warm welcome.

� We are happy to have you with us. I understand you seek a truth. Since you are an American, I am sure you want it quickly. We will try to accommodate, but please sympathize with our unhurried ways.

☦ Oh, I appreciate that. I hope I will not be much trouble.

� Not at all. I understand that you physicists just recently, in the past century, as a matter of fact, have learned some of the deeper truths of our universe. Your technology limits you to working with small and simple objects. Our "technology," if you wish to call it that, can provide a demonstration with the most complex entities.

☦ (ENTHUSIASTICALLY, BUT SUSPICIOUSLY) I'd be eager to see that.

� I have made such arrangements. You will ask an appropriate question, and the answer to your question will then be revealed to you. I believe the procedure of posing a question and having an answer revealed is much like what you scientists call "doing an experiment." Do you wish this experience?

🏃 (LOOKS PUZZLED) Why, yes I do . . . .

🛡 I will prepare a situation to allow that experiment.

The Rhob motions toward two small huts about twenty yards apart. Between the huts a young man and a young woman stand holding hands.

🛡 Arranging our situation, "preparing the state" you would call it, must be done without observation. Please don this hood.

Our visitor places the soft black hood over his head. The Rhob soon continues.

🛡 The state is now prepared. Please remove the hood. In one of these huts there is a couple, a man and a woman together. The other hut is empty. Your first "experiment" is to determine which hut holds the couple and which hut is empty. Do this by asking an appropriate question.

🏃 OK, in which hut is the couple, and which hut is empty?

🛡 Very good, well done!

The Rhob signals an apprentice, who opens the door to the right-hand hut to reveal a man and a woman arm in arm smiling shyly. He subsequently has the door of the other hut opened showing it to be empty.

🛡 Notice, my friend, you received an *appropriate* answer to your question. The couple was indeed in one of the huts. And the other hut was, of course, empty.

🏃 (UNIMPRESSED, YET TRYING TO BE POLITE) Uh huh. Yes, I see.

🛡 But I understand reproducibility is crucial to scientists. We will repeat the experiment.

Six more times this procedure is repeated for our visitor. Sometimes the couple is in the right-hand hut, sometimes in the left. Since our visitor is clearly getting bored, the Rhob stops the demonstrations and explains.

🕴 (SOMEWHAT GLEEFULLY) Notice, my friend! Your asking the where-abouts of the *couple* caused the young man and young woman to be together in a single hut.

🚶 (ANNOYED BY HAVING TRAVELED SO FAR TO SEE AN APPARENTLY TRIVIAL DEMONSTRATION, OUR VISITOR IS FINDING IT HARD NOT TO OFFEND) My *questions* caused the couple to be in one hut or the other? Nonsense! Where you placed them while I was hooded did that. Oh, but, I apologize. Thank you very much for your demonstration. But it's getting late; I must get down the mountain.

🕴 No, it is I who should apologize. I must remember that the attention span of Americans is short. I have heard that you actually choose the leaders of your nation on the basis of a number of thirty-second displays on a small glass wall.

But please, we now have a second experiment. You will ask a different question. You will ask a question causing the man and the woman to be in separate huts.

🚶 Well, yes, but I do have to be down . . .

Without waiting for our visitor to finish, the Rhob hands him the hood, and with a shrug our visitor dons it. And the Rhob speaks.

🕴 Please remove the hood. Ask a *new* question, one to determine in which hut is the man and in which hut is the woman.

🚶 OK, OK, in which hut is the man and in which hut is the woman?

*This time the Rhob signals his apprentices to open the huts **at the same time**. They reveal the man in the right-hand hut and the woman in the left, smiling at each other across the plaza.*

🕴 Notice! You received an answer appropriate to the new question you asked, a result appropriate to the *different* experiment you did. Your question caused the couple to be *distributed over both huts*. We now display reproducibility by repeating this experiment.

🚶 Please, I must be leaving. (NOW WITH A SARCASTIC TONE OF VOICE) I con-cede that your "experiments" are all repeatable an arbitrarily large number of times with equally impressive results.

🜚 Oh, I *am* sorry.

🜚 (TAKEN ABACK BY HIS OWN DISCOURTESY) Oh, no, *I* apologize. I would be delighted to see a repeat of this experiment.

🜚 Well, maybe just two or three times?

The demonstration is repeated three times.

🜚 You seem impatient. So maybe three times is enough to demonstrate that your asking the whereabouts of the man and the women *separately* caused the couple to be *spread* over both huts. Can you agree?

🜚 (BORED AND DISAPPOINTED, BUT SOMEWHAT SMUG) I surely agree that you can distribute the couple over the huts the way you wish. However, now I truly must be getting down the mountain. But thank you very much for . . .

🜚 (INTERRUPTING) You have not yet seen the *final* version of these experiments. It is the *crucial* one that completes our demonstration. Let me do it for you—just twice. Only two times.

🜚 (CONDESCENDINGLY) Well, OK, two times.

Our visitor again dons the hood.

🜚 Please remove the hood and ask your question.

🜚 Which question should I ask?

🜚 Ah, my friend, you are now experienced with both questions. You may ask either of them. You may choose either experiment.

🜚 (WITHOUT MUCH THOUGHT) OK, in which hut is the couple?

The Rhob has the door of the right-hand hut opened to reveal a man and a woman hand in hand. He then has the door of the other hut opened showing it to be empty.

🜚 (A BIT PUZZLED, BUT NOT REALLY SURPRISED) Hmmm . . . .

🜚 Notice the question you asked, the experiment you chose, caused the couple to be in a single hut. Now let's try it again—for the second time—to which you did agree.

🕺 (QUITE WILLINGLY) Sure, let's try again.

Our visitor again dons the hood.

👽 Please remove the hood and ask either question.

🕺 (WITH A TOUCH OF SKEPTICISM) OK, this time I've decided to ask the other question: In which hut is the man, and in which hut is the woman?

The Rhob has his apprentices open both huts at the same time to reveal the man in the right-hand hut and the woman in the left.

🕺 Hmmmmm. . . . (ASIDE, a spoken thought) Funny, he was able to answer the question I chose twice in a row. He could not know which one I would ask.

👽 Notice, my friend, whichever question you choose to ask is always appropriately answered. And now you wish to leave us.

🕺 Well, uh, . . . as a matter of fact, I'd like to try this last experiment again.

👽 Very well. I am delighted by your interest in the demonstration that no matter which experiment you choose, you always get an appropriate answer.

Our visitor once more dons the hood.

👽 Please remove the hood and once again, ask either question.

🕺 OK, *this* time, in which hut is the couple?

The Rhob has the door to the left-hand hut opened to reveal the man and woman together. He then has the door of the other hut opened showing it to be empty.

🕺 You had arranged an appropriate answer to the question I later chose to ask three times in a row. Your luck is impressive!

👽 It was not luck, my friend. The observation you freely chose determined whether the couple would be together in one hut or divided in two.

人 (puzzled) How can that be? (eagerly, now) Can we try that again?

♈ Surely, if you wish.

The demonstration is repeated, and our increasingly puzzled visitor requests yet further repetitions. Eight times he sees a result appropriate to the question he asked, but a result inappropriate to the other question he could have asked.

人 (AN AGITATED ASIDE: *I can't believe this!*) Please, I'd like to try this yet again!

♈ I'm afraid it now is getting dark, and it is a steep climb down the mountain. Be assured that you will always get answers appropriate to the question you ask, appropriate to the situation your question caused to exist.

人 (MUMBLES AND LOOKS BOTHERED)

♈ Something troubles you, my friend?

人 How did you know which question I was going to ask when you placed your people in the huts?

♈ I did not know. You could have asked either question.

人 (agitated) But, but . . . let's be reasonable! What if I had asked the question that did *not* correspond to where the man and woman actually were?

♈ My friend, did not your great Danish physicist, Bohr of Copenhagen, teach that science need not provide explanations for experiments not actually performed, need not answer questions not actually asked?

人 Oh yes, but come on. Your people had to be either together or separated immediately before I asked my question.

♈ I see what disturbs you. In spite of your training as a physicist, and your experience with quantum mechanics in the laboratory, you are still imbued with the notion that a particular physical reality exists before your choice of what to observe, and before your conscious experience of it. Apparently physicists find it hard to fully comprehend the great truth they have so recently gleaned. But good night, my friend. You have seen what you came to see. You must now leave us. Have a safe trip down the mountain.

🚶 (OBVIOUSLY BAFFLED AS HE TURNS TO LEAVE) Uh, yes, I will, uh, thank you very much, very much, I, uh, well . . . thank you . . .

🚶 (TALKING TO HIMSELF AS HE PICKS HIS WAY DOWN THE STEEP AND ROCKY TRAIL) Now let's see, there's got to be a reasonable explanation. If I asked where the couple was, he immediately showed the couple together in a single hut. But if I chose to ask where the man and woman each were separately, he immediately showed one of them in *each* hut. But before I asked they had to be in one situation or the other? The huts were far apart. How did he do it?!

Was I tricked into asking the question that fit the setup he had arranged? No, I *know* my choices were freely made.

It's impossible! But I saw it. It's like a quantum experiment, where both situations existed at the same time, Until you look you see only one. "Conscious experience," the Rhob said. But physics shouldn't involve anything like *consciousness!* Anyway, quantum mechanics doesn't apply to big things like people. Well . . ., of course, that's not quite right. In principle, quantum physics applies to everything. But you can only demonstrate such stuff with an interference experiment. And interference experiments are impossible with big things—for all practical purposes. Was I hallucinating?

How do I debunk this Rhob when I get back to California? And, oh my god! The guys back in the physics department will ask about my trip. Ouch!

There is, of course, no Neg Ahne Poc. What our visitor saw is in fact impossible. But in later chapters you will see how an object can be shown to be wholly in one place or, by a different choice of experiment, could have been shown to have been distributed over two locations, like the couple in Neg Ahne Poc. You should experience the same bafflement as did our Neg Ahne Poc visitor.

Demonstrating that a physical reality is caused by observation is limited by present technology to very small things. But it's being demonstrated for larger and larger objects all the time. We will devote a whole chapter to physics' "orthodox" resolution of this paradox, the Copenhagen interpretation of quantum mechanics, with Niels Bohr as its principal architect The explanation given by Bohr is not unlike that given by the Rhob in Neg Ahne Poc ("Rhob" is Bohr spelled backward.) We later discuss modern challenges to the Copenhagen interpretation.

3

Our Newtonian Worldview
A Universal Law of Motion

Nature and Nature's laws lay hid in night:
God said, Let Newton be! And all was light.

—Alexander Pope

Quantum theory conflicts violently with our intuition. Nevertheless, physicists readily accept quantum theory as the underlying basis of all physics, and thus of all science. To understand why, consider the history.

Galileo's bold stance in the seventeenth century *created* science in any modern sense of that word. And within decades, Isaac Newton's discovery of a universal law of motion became the model for all rational explanation. Newton's physics led to a worldview that today shapes the thinking of each of us. Quantum mechanics both rests on that thinking and challenges it.

Galileo insisted that a scientific theory be accepted or rejected solely on the basis of experimental test. Whether or not a theory fit one's intuition must be irrelevant. This dictum defied the scientific outlook of the Renaissance, which was, in fact, that of ancient Greece. Let's look at the problem Galileo faced in Renaissance Italy: the heritage of Greek science.

Greek Science: Its Contributions and Its Fatal Flaw

We owe the philosophers of ancient Greece credit for setting the scene for science by seeing Nature as explicable. When Aristotle's writings were rediscovered in the thirteenth century, they were revered as the wisdom of a "Golden Age."

Aristotle noted that everything that happens is essentially the motion of matter. Even, say, the sprouting of acorns to become oak trees. He therefore started by treating the motion of *simple* objects, where he could start with a small number of fundamental principles. This is indeed the way we do physics today. We search for fundamental principles. However, Aristotle's method for *choosing* fundamental principles made progress impossible. He assumed such principles could be intuitively perceived as self-evident truths.

Here are a few of them: A material object sought rest with respect to the cosmic center, which "clearly" was Earth. An object fell because of its desire for this cosmic center. A heavy object, with its greater desire, would therefore, without doubt, fall faster than a light object. In the perfect heavens, on the other hand, celestial objects moved in that most perfect of figures, the circle. These circles would be on spheres centered on the cosmic center, Earth.

Greek science had a fatal flaw: *It had no mechanism to compel consensus.* The Greeks saw experimental tests of scientific conclusions as no more relevant than experimental tests of political or aesthetic positions. Conflicting views could be argued indefinitely.

The thinkers of the Golden Age launched the scientific endeavor. However, without a method to establish some agreement, progress was impossible. Though Aristotle established no consensus in his own day, in the late Middle Ages his views became the official dogma of the Church, mostly through the effort of Thomas Aquinas.

Aquinas fitted Aristotle's cosmology and physics with the Church's moral and spiritual doctrines to create a compelling synthesis. Earth, where things fell, was also the realm of morally "fallen" man. Heaven, where things moved in perfect circles, was the realm of God and His angels. At the lowest point in the universe, at the center of Earth, was Hell. When, in the early Renaissance, Dante used this cosmological scheme in his *Divine Comedy*, it became a view that profoundly influenced Western thought.

Medieval and Renaissance Astronomy

The position of the stars in the sky foretold the change of the seasons. What, then, was the significance of the five bright objects that wandered through the starry background? An "obvious" conclusion was that the motion of these planets ("planet" means wanderer) foretold erratic human affairs. The planets therefore warranted serious attention. Astronomy's roots are in astrology.

In the second century A.D., Ptolemy of Alexandria mathematically described the heavenly motions so well that calendars and navigation based on his model worked beautifully. The astrologer's predictions—at least regarding the positions of the planets—were likewise accurate. Ptolemy's astronomy, with a stationary Earth as the cosmic center, required planets to move on "epicycles," complicated loopy curves made up of circles rolling on circles within yet further circles. King Alfonso X of Castile, having the Ptolemaic system explained to him, supposedly remarked: "If the Lord God Almighty had consulted me before embarking on Creation, I would have recommended something simpler." Nevertheless, the combination of Aristotle's physics with Ptolemy's astronomy was accepted as both practical truth and religious doctrine, and was enforced by the Holy Inquisition.

Then, in the sixteenth century, an insight upsetting the whole apple cart appeared within the Church itself. The Polish cleric and astronomer Nicolas Copernicus felt Nature had to be simpler than Ptolemy's cosmology. He suggested that Earth and five other planets orbited a central, stationary sun. The back-and-forth wandering of the planets against the starry background was a result of our observation of them from an also-orbiting Earth. Earth was just the third planet from the sun. It was a simpler picture.

Simplicity was hardly a compelling argument. Earth "obviously" stood still. One *felt* no motion. A dropped stone would be left behind on a moving Earth. Since air was presumed to occupy all space, if Earth moved, a great wind would blow. Moreover, a moving Earth conflicted with the wisdom of the Golden Age. Such arguments were hard to refute. And, most disturbingly, the Copernican system was seen to contradict the Bible, and doubting the Bible threatened salvation.

Copernicus's work, published shortly before his death, included a foreword added by an editor declaring his description to be a mathematical convenience only. It did not presume to describe *actual* motions. Any contradiction of the Church's teachings was disavowed.

A brilliant analysis some decades later by Johannes Kepler showed that accurate new data on the motion of the planets fit perfectly if he assumed that planets moved in *elliptic* orbits with the sun at one focus. He also discovered a simple rule giving the exact time it took each planet to orbit the sun, depending on its distance from the sun. Kepler could not explain his rule, and he disliked ellipses, "imperfect circles," but rising above prejudice, he accepted what he saw.

Kepler did great astronomy, but science did not guide his contemporary worldview. Initially, he considered the planets to be pushed along their orbits by angels, and as a sideline he drew horoscopes, in which he likely believed. He also had to take time from his astronomy to defend his mother from accusations of witchcraft.

Galileo's New Ideas on Motion

In 1591, only twenty-seven years old, Galileo became a professor at the University in Padua, but he soon left for a post at Florence. Today's university faculty would understand why: more time for research and less teaching. Galileo's talents included music and art as well as science. Brilliant, witty, and charming, he could also be arrogant, brash, and petty. We envy his skill with words. He liked women, and they liked him.

Galileo was a convinced Copernican. That simpler system made sense to him. But unlike Copernicus, Galileo did not merely claim a new technique for calculation; he argued for a new worldview. A humble approach was not his style.

The Church had to stop Galileo's call for independent thought. Saving souls was the Church's goal, not scientific validity. Found guilty of heresy by the Holy Inquisition, and given a tour of the torture chambers, Galileo recanted his claim of a sun-orbiting Earth. For his last years, Galileo lived under house arrest—a lesser penalty than that of another Copernican, Giordano Bruno, who was burned at the stake.

Recantation notwithstanding, Galileo knew that Earth moved. Moreover, he realized that Aristotle's explanation of motion could not survive on a moving Earth. Friction, not desire for rest in the cosmic center, caused a sliding block to stop. Air resistance, not less desire for the cosmic center, caused a feather to fall more slowly than a stone.

Contradicting Aristotle's claims, Galileo asserted: "In the absence of friction or other impressed force, an object will continue to move

Figure 3.1 Galileo Galilei. © National Maritime Museum, Greenwich, London

horizontally at a constant rate." And: "In the absence of air resistance heavy objects and light objects will fall at the same rate."

Galileo's ideas were obvious—*to him*. How could he convince others? Rejecting Aristotle's teaching on the motion of matter was not a minor issue. Aristotle's philosophy was an all-encompassing, Church-enshrined worldview. Reject a part, and you appear to reject it all.

The Experimental Method

To compel agreement with his ideas, Galileo needed examples that conflicted with Aristotle's mechanics, but examples that conformed to his own ideas. Looking around, he could see few such cases. His solution: *create them!*

Galileo would contrive special *clear-cut* situations: "experiments." An experiment tests a theoretical prediction. This may seem an obvious approach, but in that day it was an original and profound idea.

In his most famous experiment, Galileo supposedly dropped a ball of lead and a ball of wood from the leaning Tower of Pisa. The simultaneous click of the wood and the thud of the lead proved the light wood fell as fast as the heavy lead. Such demonstrations gave reason enough, he argued, to abandon Aristotle's theory and to accept his own.

Some faulted Galileo's experimental method. Though the displayed facts could not be denied, Galileo's demonstrations were "merely *contrived* situations." They could be ignored because they conflicted with the intuitively obvious nature of matter. Moreover, Galileo's ideas *had* to be wrong because they conflicted with Aristotelian philosophy.

Galileo had a far-reaching answer: Science should deal only with those matters that can be demonstrated. Intuition and authority have no standing in science. *The only criterion for judgment in science is experimental demonstration.*

Within a few decades, Galileo's approach was accepted with a vengeance. Science progressed with a vigor never before seen.

Reliable Science

Let's agree on some rules of evidence for accepting a theory as reliable science. They will stand us in good stead when we consider the acceptance of counterintuitive quantum theory.

But first, a remark on the word "theory": We speak of quantum *theory* but of Newton's *laws*. "Theory" is the modern word. We can't think of a single twentieth- or twenty-first-century "law" in physics. Though "theory" is sometimes used for a speculative idea, it does not necessarily imply uncertainty. Quantum *theory* is, as far as is known, completely correct. Newton's *laws* are an approximation.

For a theory to compel consensus, it must, first of all, make predictions that are testable, with results that can be displayed objectively. It must stand with a chip on its shoulder challenging would-be refuters.

"If you're good, you'll go to Heaven." That prediction may well be correct, but it's not objectively testable. Religions, political stances, or philosophies in general are not scientific theories. Aristotle's testable theory of falling, predicting that a two-pound stone will fall twice as fast as a one-pound stone, is a scientific theory, albeit a wrong one.

A theory making testable predictions is a *candidate* for being reliable science. Its predictions must be tested by experiments that challenge the theory by attempting to refute it. And the experiments must be convincing even to skeptics. For example, theories suggesting the existence of extra-sensory perception (ESP) make predictions, but so far, tests have not been convincing to skeptics.

To qualify as reliable science, a theory must have many of its predictions confirmed without a single disconfirmation. A single incorrect prediction forces a theory's modification or abandonment. The scientific method is hard on theories. One strike and you're out! Actually, no scientific theory is ever *totally* reliable. It is always possible that it will fail some future test. A scientific theory is, at best, *tentatively* reliable.

The scientific method, setting high standards for experimental verification, is hard on theories. But it can also be hard on us. If a theory meets these high standards, we are obligated to accept it as reliable science, no matter how violently it conflicts with our intuitions. Quantum theory will be our case in point.

The Newtonian Worldview

Isaac Newton was born in 1642, the year Galileo died. With the wide acceptance of the experimental method, there was a sense of scientific progress, though Aristotle's erroneous physics was still often taught. The Royal Society of London, today a major scientific organization, was founded in 1660. Its motto, *Nullis in verba*, translates loosely as, "Take nobody's word for it." It would have delighted Galileo.

Newton, a handy fellow, was supposed to take over the family farm. But more interested in books than plows, he managed to go to Cambridge University, paying his way by working at menial tasks. He did not shine as a student, but science fascinated him—"natural philosophy," it was then called. When the Great Plague forced the university to close, Newton returned to the farm for a year and a half.

Young Newton understood Galileo's teaching that on a perfectly smooth horizontal surface a block, once moving, would slide forever.

Figure 3.2 Portrait of Isaac Newton (1642–1727) 1702.
Photograph by Sir Godfrey Kneller, courtesy of Getty Images

A force is needed only to overcome friction. With a greater force, the block would speed up. It would accelerate.

Galileo, however, had accepted the Aristotelian concept that falling was "natural" and needed no force. He also assumed the planets moved "naturally" in circles without any force. Galileo just ignored the ellipses discovered by his contemporary, Kepler. For Newton to conceive his universal laws of motion and gravity, he had to move beyond Galileo's acceptance of Aristotelian "naturalness."

Newton tells that his inspiration came as he watched an apple fall. He likely asked himself: Since a force was needed for *horizontal* acceleration, why not a force for *vertical* acceleration? And if there's a downward force on an apple, why not on the moon? If so, why doesn't the moon fall to Earth like the apple?

In Newton's famous cannon-on-a-mountain sketch, the dropped cannonball falls straight downward, while those fired with larger velocities land farther away. If a ball is fired fast enough, it will miss the planet. It nevertheless continues to "fall." It continues to accelerate toward Earth's center while it also moves "horizontally." It thus orbits Earth. As the cannon ball comes around, the cannoneer had better duck.

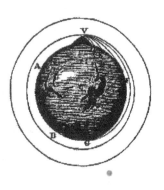

Figure 3.3 Newton's drawing of a cannon on a mountain

The moon doesn't crash to Earth only because it, like that fast cannon ball, has a velocity perpendicular to Earth's radius. Newton realized what no one had before: The moon *is* falling.

The Universal Law of Motion and, *Simultaneously*, a Force of Gravity

Galileo thought that uniform motion without force applied only to motion that was parallel to the surface of Earth, in a *circle* about Earth's center. Newton corrected this to say that a force is needed to make a body deviate from a constant speed in a *straight line*.

How much force is needed? The more massive the body, the more force should be needed to accelerate it. Newton speculated that the force needed was just the mass of the body times the acceleration that force produced, or $F = Ma$. It's Newton's universal law of motion.

In Newton's day, however, there seemed to be a counterexample: Falling was a downward acceleration, but apparently *without* an impressed force. Young Newton had to conceive two profound ideas *simultaneously*: his law of motion *and* the force of gravity.

When the plague subsided, Newton returned to Cambridge. Isaac Barrow, then Lucasian Professor of Mathematics, was soon so impressed with his one-time student that he resigned to allow Newton to take the Lucasian chair. The quiet boy became a reclusive bachelor. (Celibacy was then required of Cambridge faculty.) Newton was reserved and moody and was often angered by well-intended criticism. You'd rather spend the evening with Galileo.

Newton's ideas needed testing. However, the force of gravity between objects that he could move about on Earth was far too small for him to measure. So he looked to the heavens. Using his equation of motion and his law of gravity, he derived a simple formula. A chill no doubt ran down his spine when he saw it. His formula was precisely the unexplained rule Kepler had noted decades earlier for the exact time it took each planet to orbit the sun.

Newton could also calculate that the twenty-seven-day orbital period of the moon was just what you would expect if a falling object gained a speed of ten meters per second in each second of falling, just the acceleration experimentally demonstrated by Galileo. Newton's equations of motion and gravity governed apples as well as the moon. It is on Earth as it is in the heavens. Newton's equations are universal.

Principia

Newton realized the significance of his discoveries, but controversy over the first paper he wrote had upset him. The idea of publishing now terrified him.

Some twenty years after his insights back on the farm, Newton was visited by the young astronomer Edmund Halley. Knowing others were speculating on a law of gravity that would yield Kepler's elliptical orbits for the planets, Halley asked Newton what orbits his law of gravity would predict. Newton immediately answered, "ellipses." Impressed by the quick response, Halley asked to see the calculations. Newton could not find his papers. A historian notes: "While others were still seeking a law of gravity, Newton had already lost it."

After Halley's warning that others might scoop him, Newton spent a furious eighteen months producing *Philosophiae Naturalis Principia Mathematica*. What we now just call *Principia* was published in 1687 with Halley footing the bill. Newton's fears of criticism were realized; some even claimed he stole their work.

Though *Principia* was widely recognized as the profound revelation of Nature's laws, being mathematically rigorous and in Latin, it was little read.

But popularized versions soon appeared. *Newtonianism for Ladies* was a best-seller. Voltaire, aided by his more scientifically talented companion, Madame du Châtelet, in his *Elements of Newton* claimed to "reduce this giant to the measure of the nincompoops who are my colleagues."

The revealed rationality of Nature was revolutionary. It implied that, in principle at least, the world should be as understandable as the mechanism of clocks. That clockwork aspect was later dramatically demonstrated by Halley's accurate prediction of the return of a comet. Until then, a comet was commonly thought to foretell the death of a king.

Principia ignited the intellectual movement known as the Enlightenment. Society would no longer look to the Golden Age of Greece for wisdom. Alexander Pope captured the mood: "Nature and Nature's laws lay hid in night: / God said, Let Newton be! And all was light."

When he needed better mathematics, Newton invented calculus. His studies of light transformed the field of optics. He held the chair in Parliament then reserved for Cambridge. He became Director of the Mint and took the position seriously. In his later years, Sir Isaac—the first scientist ever knighted—was perhaps the most respected person in the Western world. Paradoxically, Newton was also a mystic, immersing himself in supernatural alchemy and the interpretation of Biblical prophecies.

Newton's Legacy

The most immediate impact of the Newtonian worldview was to break the late-medieval synthesis of the physical and the spiritual. While Copernicus had, unintentionally perhaps, initiated the destruction of this Church-sponsored relationship by denying Earth as the cosmic center, Newton completed the job by showing that the same *physical* laws held for both the earthly and the heavenly realms. Under this inspiration, geologists, assuming that the same laws also applied throughout time, showed Earth to be vastly older than the Bible's 6,000 years. This led directly to Darwin's theory of evolution, the most socially disturbing idea of modern science.

Though aspects of Newton's legacy will forever endure, the Newtonian mechanistic worldview, and what we today call "classical physics," is challenged by modern physics. But the mechanistic worldview, our Newtonian heritage, still molds our commonsense view of the physical world and shapes our thinking in every intellectual sphere.

We now focus on five "commonsense" Newtonian stances. Quantum mechanics challenges each of them.

Determinism

Idealized billiard balls are the physicist's much-loved model for determinism. If you know the position and velocity of a pair about to collide, with Newton's physics you can predict their position and velocity arbitrarily far into the future. Computers can calculate the future positions of a large number of colliding balls.

The same might be said, in principle, for the atoms bouncing around in a box of gas. Take this idea all the way: To an "all-seeing eye" that knew the position and velocity of each atom in the universe at a given moment, the entire future of the universe would be apparent. The future of such a Newtonian universe is, in principle, *determined*—whether or not anyone *knows* that future. The deterministic Newtonian universe is the Great Machine. The meshing gears of its clockworks move it on a predetermined course.

God then becomes the Master Clocksmith, the Great Engineer. Some went further: After making the completely deterministic machine, God had no role. He was a *retired* engineer. Moving from retirement to nonexistence was a small step.

Determinism gets personal: Are your seemingly free choices actually predetermined? According to Isaac Bashevis Singer, "You have to believe in free will. You have no choice." We have a paradox: Our perception of free will conflicts with the determinism of Newtonian physics.

What about free will *before* Newton? No problem. In Aristotle's physics even a stone followed its individual inclination as it rolled down the hill in its own particular way. It is the determinism of Newtonian physics that presents the paradox.

It is, however, a benign paradox. Though we affect the physical world by our conscious free will, the externally observable effects of free will on the physical world come about indirectly, through muscles that physically move things. Consciousness itself can be seen as confined within our body.

Classical physics thus allows the tacit isolation of consciousness and its associated free will from the domain of the physicist's concern. There is mind, and then there is matter. Physics deals with matter. With this divided universe, prequantum physicists could logically avoid the paradox. The determinism/free will paradox could be avoided because it arose only through the deterministic *theory*, not through any experimental demonstration. Thus, by limiting the scope of the theory to exclude the observer, physicists could relegate free will and the rest of consciousness to psychology, philosophy, and theology. And that was their inclination.

Determinism was challenged at the inception of quantum mechanics, when Max Planck had electrons behaving randomly. A later, more profound challenge, was the intrusion of the observer into the quantum *experiment*. No longer can the issue of free will be simply ruled out of physics by limiting the scope of the theory. It arises in the *experimental demonstration*. With quantum mechanics, the paradox of conscious free will is no longer benign.

Physical Reality

Before Newton, explanations were mystical, and largely useless. If planets were pushed by angels, and rocks fell because of their innate desire for the cosmic center, if seeds sprouted by craving to emulate mature plants, who could deny the influence of other occult forces? Or that the phases of the moon, or incantations, might be relevant? The flu, its full name "influenza," is so named because it was originally explained in terms of a supernatural *influence*.

By contrast, in the Newtonian worldview, Nature was a machine whose workings, though incompletely understood, need be no more mysterious than the clock whose gears are not seen. Acceptance of such a physically real world has become our conventional wisdom. Though we may say the car "doesn't want to start," we expect the mechanic to find a physical explanation.

We raise the issue of "reality" because quantum mechanics challenges this classical view of it. But let's avoid a semantic misunderstanding. We're *not* talking of *subjective* reality, a reality that can differ from one person to the next. For example, we may say, "You create your own reality," meaning your *psychological* reality. The reality we're talking of here is *objective* reality, a reality we can all agree on, like that of a rock's position.

For millennia philosophers have taken widely varied stands on the nature of reality. A conventional philosophical stance called "realism" has the existence of the physical world being independent of its observation. A more drastic version denies the existence of anything *beyond* physical objects. In this "materialist" view, consciousness, for example, should be ultimately understandable, in principle at least, in terms of the electro-chemical properties of the brain. The tacit acceptance of such a materialist view, even its explicit defense, is not uncommon today.

Contrasting with Newtonian realism or materialism is the philosophical stance of "idealism" holding that the world that we perceive is not the *actual* world. Nevertheless, the actual world can be grasped with the mind.

An extreme idealist position is "solipsism." Here's its essence: *All* I ever experience are my own sensations. All I can know of my pencil, for example, is the sensation of its reflected light on my retina and its pressure against my fingers. I cannot demonstrate that there is anything "real" about the pencil, or about anything else, beyond my experienced sensations. (Appreciate that this paragraph is in first-person singular. The rest of you are, solipsistically speaking, just sensations in my mental world.)

"If a tree falls in the forest, and no one hears it, is there any noise?" The realist answers: "Yes, even if the air pressure variations we might experience as sound were heard by no one, they existed as a physically real phenomenon." The solipsist answers: "No, there wasn't even a tree unless I experienced it. Even then, only my conscious sensations *actually* existed." In this regard, we quote philosopher Woody Allen: "What if everything is an illusion, and nothing exists? In that case, I've definitely overpaid for my carpet."

We'll see that the intrusion of the conscious observer into the quantum experiment jolts our Newtonian worldview so dramatically that the philosophical issues of realism, materialism, idealism—even silly solipsism—come up for discussion.

Separability

Renaissance science with its Aristotelian basis was replete with mysterious connectivities. Stones had an eagerness for the cosmic center.

Acorns sought to emulate nearby oaks. Alchemists believed their personal purity influenced the chemical reactions in their flasks. By contrast, in the Newtonian worldview, a hunk of matter, a planet or a person, interacts with the rest of the world *only* through the physically real forces impressed upon it by other objects. It is otherwise *separable* from the rest of the universe. In this view, except for impressed physical forces, an object has no "connectedness" with the rest of the universe.

Physical forces can be subtle. For example, when a fellow, seeing a friend, adjusts his motion to meet her, the influencing force is carried by the light reflected from his friend and is exerted on rhodopsin molecules in his retina. An example of a *violation* of separability would be a voodoo priest sticking a pin in a doll and thereby, without a connecting physical force, cause you pain.

Violating our classical intuition, quantum mechanics includes instantaneous influences that *violate* separability. Einstein derided these as "spooky actions." However, actual experiments now demonstrate that they do in fact exist.

Reduction

Often implicit in viewing the world as comprehensible is the reductionist hypothesis that a complex system can, in principle at least, be explained in terms of, or "reduced" to, its simpler parts. The working of an automobile engine, for example, can be explained in terms of the pressure of the burning gasoline pushing on the pistons.

Explaining a psychological phenomenon in terms of its biological basis would be reducing an aspect of psychology to biology. ("There is more of gravy than of grave to you," said Scrooge to Marley's ghost as he reduced his dream to a digestive problem.)

A chemist might explain a chemical reaction in terms of the physical properties of the involved atoms, something feasible today in simple cases. This would be reducing a chemical phenomenon to physics.

We can think of a reductionist pyramid, a hierarchy going from psychology to physics, with physics being firmly based on empirical facts. Scientific explanations are generally reductionist, moving toward more basic principles. Though one moves in that *direction*, it is usually only by small steps. We will always need general principles specific to each level.

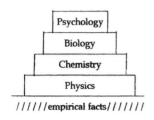

Figure 3.4 Hierarchy of scientific explanation

The classic example of a violation of reductionism is the "vital force" once proposed to account for life processes. Life, in this view, emerged at the biological level without an origin in chemistry or physics. Such vitalist thinking led nowhere and, of course, has no standing in today's biology.

In studies of consciousness, reduction sparks controversy today. Some argue that once the electrochemical neural correlates of consciousness are understood, there will be nothing left to explain. Others insist that the "inner light" of our conscious experience will elude the reductionist grasp, that consciousness is primary, and that new "psychophysical principles" will be needed. Quantum mechanics is claimed as evidence supporting this non-reductionist view.

A Sufficient Explanation

Newton was challenged to *explain* his force of gravity. A force transmitted through empty space, through *nothingness*, was a big pill to swallow.

Newton had a succinct response: "*Hypotheses non fingo*" ("I make no hypotheses"). He thus claimed that a theory need do no more than provide consistently correct predictions. The *hypotheses non fingo* attitude arises again with quantum mechanics. However, quantum theory's denial of a straightforward physical reality is an even bigger pill to swallow than a force transmitted through nothingness.

Beyond Physics by Analogy

In the decades following Newton, engineers learned to build the machines that launched the Industrial Revolution. Chemists moved beyond mystical alchemy, which for centuries had achieved almost nothing. Agriculture became scientific as understanding replaced folklore. Though the early workers in technology used almost no physics, the rapid advances they made required the Newtonian perspective that discernable laws govern the physical world.

Newton's physics became the paradigm for all intellectual endeavors. Analogies with physics were big and bold. Auguste Comte, inventing the term "sociology," referred to it as "social physics," in which people were "social atoms" motivated by forces. The study of society had never previously been regarded as scientific.

Pushing the analogy with Newtonian physics, Adam Smith argued for laissez-faire capitalism by claiming that if people were allowed to pursue their own interests, an "invisible hand," a fundamental law of political economy, would regulate society for the general good.

Analogies are flexible. Karl Marx felt that he, not Adam Smith, had discovered the correct law. Marx claimed to "lay bare the economic law of motion." With that law he predicted the communist future. By analogy with a mechanical system, he merely needed to know the initial condition, which, he thought, was the capitalism of his day. Thus, Marx's major work, *Das Kapital*, is a study of capitalism.

Analogies also arose in psychology. Sigmund Freud wrote: "It is the intention of this project to furnish us with a psychology which shall be a natural science. Its aim is to represent psychical processes as quantitatively determined states of specific material particles . . . ." Newtonian enough? As a later example, consider B. F. Skinner's declaration: "The hypothesis that man is not free is essential to the application of the scientific method to the study of human behavior." He explicitly denies free will, polemically adopting materialism and Newtonian determinism.

The appeal of such approaches in the social sciences has cooled. Workers in such complex areas are today more aware of the limitations of a method that works well for simpler physical situations. But the *broader* Newtonian perspective, the seeking of general principles that are then subject to empirical tests, is the accepted mode.

The Newtonian perspective is our intellectual heritage. We can hardly escape it. It is the basis of our everyday common sense. Even our *scientific* commonsense. Being explicit about it can help us appreciate the challenge that quantum mechanics poses to that classical worldview.

4

All the Rest of Classical Physics

There is nothing new to be discovered in physics now.
All that remains is more and more precise
measurement.

—Lord Kelvin (in 1894)

In 1900, six years after he made this claim, Kelvin hedged: "Physics is essentially complete: There are just two dark clouds on the horizon." He picked the right clouds. One hid relativity, the other, quantum mechanics. But before we look behind those clouds, we tell a bit more of the nineteenth-century physics we today call "classical." We describe "interference," the phenomenon that demonstrates something to be an extended wave. We develop the concept of electric field, because light is a rapidly varying electric field, and it is with light that the quantum enigma first arose. We talk of energy and its "conservation," its unchanging totality. And, finally, we will tell of Einstein's theory of relativity. Accepting relativity's hard-to-believe, but well-confirmed predictions are good psychological practice for the impossible-to-believe implications of quantum theory. There's more in this chapter than you actually *need* to know in order to understand the quantum enigma. But it's good background.

The Story of Light

Newton decided that light was a stream of tiny particles. He had good arguments. Just like objects obeying his universal equation of motion, light travels in straight lines unless it encounters something that might exert a force on it. In Newton's words: "Are not the Rays of Light very small Bodies emitted from shining Substances? For such Bodies will pass through uniform Mediums in right Lines without bending into the Shadow, which is the Nature of the Rays of Light."

Actually, Newton was conflicted. He investigated a property of light we now call "interference," a phenomenon uniquely characteristic of extended waves. Nevertheless, he came down strongly in favor of particles. His reasoning was that waves would require a medium in which to propagate, and this medium would impede the motion of the planets, something his universal equation of motion seemed to deny. As he put it:

> And against filling the Heavens with fluid Mediums, unless they be exceeding rare, a great Objection arises from the regular and very lasting Motions of the Planets and Comets in all manner of Courses through the Heavens. . . . [T]he Motions of the Planets and Comets being better explain'd without it. . . . [S]o there is no evidence for its Existence, and therefore it ought to be rejected. And if it be rejected, the Hypotheses that Light consists in Pression or Motion, propagated through such a Medium, are rejected with it.

Other scientists proposed wave theories of light, but the overwhelming authority of Newton meant that his "corpuscular theory," that light is a hail of little bodies, dominated for more than 100 years. The Newtonians were in fact more sure of Newton's corpuscles than was Newton, until about 1800, when Thomas Young showed otherwise.

Young was a precocious child who, reportedly, read fluently at the age of two. He was educated in medicine, earned his living as a physician, and was an outstanding translator of hieroglyphics. But his main interest was physics. At the beginning of the nineteenth century, Young provided the convincing demonstration that light was a wave.

On a glass plate made opaque with soot, Young scribed two closely spaced parallel lines. Light shining through these two slits falling onto a wall produced bright and dark bands we call an "interference pattern." We'll see that such a pattern demonstrates light to be a spread-out wave.

We can picture a "wave" as a moving series of peaks and valleys, or crests and troughs. Such crests and troughs can, for example, be seen through the flat side of an aquarium as ripples on the water surface. Another way to depict waves is the bird's-eye view, where we draw lines to indicate the crests. Waves on the ocean seen from an airplane look like this. We'll display waves both ways, as in figure 4.1.

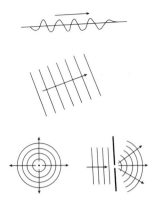

Waves from a small source, say from a pebble dropped into water, spread in all directions. Similarly, light from a small glowing object spreads in all directions.

Figure 4.1 Views of waves

By the same token, light coming from a small source, say coming through a narrow slit, will spread in all directions, and falling on a screen illuminates the screen rather uniformly. (The diagram views the slit edgewise.)

Light coming through *two* closely spaced slits might be expected to illuminate the screen twice as brightly as light from a single slit. That would certainly be expected if light were a stream of little particles, Newton's corpuscles. But when Thomas Young shined light through his two slits, he saw bands of brightness and darkness. And, most crucially, *the distance between the bright and dark bands depended on the spacing of the slits.* A stream of independent particles, each coming through a single slit, could not account for such behavior.

Interference is central to the quantum theory and to the quantum enigma, and in the next few paragraphs we explain it in a bit more detail. Interference is accepted in physics as the conclusive demonstration of extended wave behavior. You might want to just go along with that acceptance and merely skim the next several paragraphs explaining interference. Skimming down to the section titled "The Electromagnetic Force," you will still be able to appreciate the quantum enigma.

Here's how interference comes about: At the central place on the screen (point A in figure 4.2), light waves from the top slit travel exactly the same distance as do waves from the bottom slit. Therefore, crests from one slit arrive together with crests from the other. And troughs from both slits come to point A at the same time. Identical waves from the two slits arriving at A add to produce more brightness than would exist if only one slit were open.

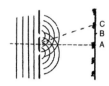

Figure 4.2 Interference
in the double-slit
experiment

But to reach a place above the central location on the screen (say point B in figure 4.2), waves from the bottom slit must travel a bit farther than waves from the top slit. Therefore, at point B, crests from the bottom slit arrive *later* than crests from the top slit. In particular, at point B crests from the bottom slit arrive at the same time as *troughs* from the top slit. These crests and troughs arriving together cancel each other. At point B the waves from the two slits *subtract* to produce dark. Light combined with other light can produce dark.

At a place yet farther up the screen (point C in figure 4.2), there will be another bright band, because at that place, once again, crests from one slit arrive with crests from the other. Continuing up the screen, bright and dark bands will alternate as waves from the two slits alternately reinforce and cancel each other to form the interference pattern. "Interference" is actually a misnomer. Waves from the two slits do not *interfere* with each other; they just add and subtract, like deposits and withdrawals from your bank account.

We are assuming that the waves coming to the slits all have the same frequency, the same distance between crests, the same wavelength. That is, we assume that the light is all of a single color. Were that not so, different colors would have their bright bands at different places, resulting in a blurred interference pattern.

If you think about the geometry, you can see that the greater the spacing between the slits, the smaller is the spacing between the bright bands of the interference pattern. The details here are not crucial. The important thing to remember is that *the pattern spacing depends on the slit spacing*. Young argued that since the amount of light at each point on the screen depends on the slit spacing, each point on the screen received light from *both* slits.

Were light a stream of particles, there would, presumably, be no interference pattern. Little bullets, independent of each other, each coming through *one* slit or the other, could not cancel each other to produce a pattern dependent on the slit separation.

Is Young's argument airtight? Probably not. When Young presented it, it was hotly disputed. Young's English colleagues were strong in the Newtonian particle school of thought. Light waves were favored by *French* scientists and were rejected by the English partly for *just* that reason. Nevertheless, further experiments soon overwhelmed objections, and light was accepted to be a wave.

We described interference in terms of light waves. But our discussion applies to waves of any kind. Our crucial point: Interference demonstrates an entity to be a spread-out wave. Interference *cannot* be explained by a stream of compact, independent objects.

The Electromagnetic Force

A piece of silk that has been rubbed on glass is then attracted to the glass. But it is repelled by another piece of silk that was rubbed on glass. Such forces due to "electric charge," seen when different materials were rubbed together, were long known. A crucial step in understanding it was the bright idea of Benjamin Franklin. He noticed that when any two electrically charged *attracting* bodies came into contact, their attraction lessened. Not so for charged bodies which repelled each other. He realized that those bodies that *attracted* each other *canceled* each other's charge.

Cancellation is a property of positive and negative numbers. Franklin therefore assigned algebraic signs, positive (+) and negative (−), to charged objects. Bodies with charges of opposite sign attract each other. Bodies with charges of the same sign repel each other.

(Franklin's work on electricity is in good part responsible for the existence of the United States. As ambassador to France, it was not just Franklin's wit, charm, and political acumen but also his stature as a scientist that allowed him to recruit the French aid that was so crucial to the success of the American Revolution.)

We now know that atoms have a positively charged nucleus made up of positively charged protons (and uncharged neutrons). Electrons, each with a negative charge equal in magnitude to the positive charge of a proton, surround the nucleus. The number of electrons in an atom is equal to the number of protons, so the atom as a whole is

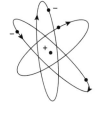

Figure 4.3 Positive and negative charges

uncharged. When two bodies are rubbed together, it is the electrons that move from one to the other.

A glass rod that is rubbed with a silk cloth, for example, becomes positively charged because electrons in the glass are less tightly bound than those in the silk. Therefore, some electrons move from the glass to the silk. The silk, now having more electrons than protons, is negatively charged and is attracted to the positively charged glass. Two negatively charged pieces of silk would repel each other.

A simple formula, Coulomb's law, tells us the strength of the electric force that one charged body (or "charge") exerts on another. With it you can calculate the forces in any arrangement of charges. That was the whole story of electric force. There was nothing more to say, or so thought most physicists in the early nineteenth century.

Figure 4.4 Michael Faraday. Courtesy Stockton Press

Michael Faraday, however, found the electric force puzzling. Let's back up a bit. In 1805, at the age of fourteen, Faraday, the son of a blacksmith, was apprenticed to a bookbinder. A curious fellow, Faraday was fascinated by some popular science lectures by Sir Humphrey Davy. He took careful notes, bound them into a book, presented them to Sir Humphrey, and asked for a job in his laboratory. Though hired as a menial assistant, Faraday was soon allowed to try some experiments of his own.

How, Faraday wondered, could one body cause a force on another through empty space? Merely that the mathematics of Coulomb's law correctly predicted what you would observe was not a sufficient explanation for him. (No "*Hypothesis non fingo*" for him.) Faraday postulated that an electric charge creates an electric "field" in the space around itself, and it is this physical field that exerts the forces on other charges. Faraday represented his field by continuous lines emanating from a positive charge and going into a negative charge. Where the lines were most dense, the field would be greatest.

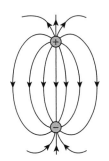

Figure 4.5 Electric field around two charges

Most scientists, claiming that the math of Coulomb's law was the whole story, considered Faraday's field concept superfluous. Faraday's ignorance of mathematics, they noted, required him to think in pictures. Abstract thinking was presumed difficult for this young man from the "lower classes." The field concept was ridiculed as "Faraday's mental crutch."

Actually, Faraday went further and assumed that the field due to a charge takes time to propagate. If, for example, a positive and a nearby negative charge of equal magnitude were brought together to cancel each other, the field would disappear in their immediate neighborhood. But it seemed unlikely to Faraday that the field would disappear everywhere immediately.

He thought the remote field would exist for a while even after the charges that created it canceled each other and no longer existed. If this were true, the field would be a physically real thing in its own right.

Moreover, Faraday reasoned, if two equal and opposite charges were repeatedly brought together and separated, an alternating electric field would propagate from this oscillating pair. Even if they stopped oscillating and just canceled each other, the oscillating field would continue to propagate outward.

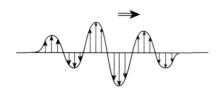

Figure 4.6 An oscillating electric field

Faraday's intuition was sound. A few years later James Clerk Maxwell, picking up Faraday's field idea, devised a set of four equations involving fields that encompassed all electric and magnetic phenomena. We call them "Maxwell's equations." They predicted the existence of waves of electric and magnetic field: "electromagnetic waves." Maxwell noticed that the speed of such waves was exactly the speed measured for light. He therefore proposed that light was an electromagnetic wave. This was soon demonstrated, unfortunately after his death.

As Faraday had predicted, the back and forth motion of the equal and opposite charges, actually *any* acceleration of electric charge, produces electromagnetic radiation. The frequency of the motion of the charges (the repeats per second) is the frequency of the wave produced. Higher frequency motion produces violet and ultraviolet light; lower frequencies produce red and infrared. Yet higher frequencies produce x-rays, much lower frequencies produce radio waves.

Today, the fundamental theories of physics are all formulated in terms of fields. Faraday's "mental crutch" is a pillar upon which all of physics now rests.

The electric force, short for electromagnetic force, is the only force we will need to talk of. Along with gravity, it is the only force we normally experience. (Though all bodies exert gravitational forces on each other, gravity is only significant when at least one of the bodies is very massive, such as a planet.) The forces between atoms are electrical.

When we touch someone, the pressure of our touch is an electric force. The electrons in the atoms of our hand repel the electrons in the atoms of the other person. Talk to someone by telephone, and it's the electric force that carries the message over wires, in optical fibers, and through space. The atoms making up solid matter are held together by electric forces. Electric forces are responsible for all of chemistry and therefore underlie all biology. We see, hear, smell, taste, and touch with electric forces. The neurological processes in our brains are electrochemical, and therefore ultimately electrical.

Is our thinking, our consciousness, ultimately to be explained wholly in terms of the electrochemistry taking place in our brain? Is our feeling of being conscious "merely" a manifestation of electrical forces? Some believe so. Others claim there is more to consciousness than electrochemistry. It's an issue we explore later, and quantum mechanics is relevant.

There are forces in nature besides gravity and the electromagnetic force. But, it seems, there are only two others: the so-called "strong force" and the "weak force." They both involve interactions of the particles making up the atomic nucleus (and objects created in high-energy particle collisions and lasting for only an instant). They exert essentially no effect beyond the dimensions of the atomic nucleus. They'll not be important in this book.

Energy

Energy is a concept pervading physics, chemistry, biology, and geology, as well as technology and economics. Wars have been fought over the chemical energy stored in oil. The crucial aspect of energy is that, though its form may change, the *total* amount of energy stays constant. That fact is called the "conservation of energy." But what *is* energy? The best way to define energy is to point to several of its different forms.

First of all, there is energy of motion. The larger the mass and the speed of a moving object, the larger its "kinetic energy." Energy due to the motion of objects is kinetic energy.

The farther a rock falls, the faster it goes and the larger its kinetic energy. A rock held at a certain height has the *potential* of gaining a certain speed, a certain kinetic energy. It has a gravitational "potential energy," which is larger for a larger mass or a larger height. The sum of a rock's kinetic and potential energy, its *total* energy, remains constant as the rock falls. This is an example of the conservation of energy.

Of course, after the rock hits the ground, it has zero kinetic energy and zero potential energy. As it contacts the ground, the energy of the rock itself was not conserved. But the *total* energy is conserved. On impact, the rock's energy is given to the random motion of the atoms of the ground and those of the rock. Those atoms now jiggle about with greater agitation. The haphazard motion of these atoms is the microscopic description of thermal energy (heat). Where the rock hit, the ground is warmer. The energy imparted to the jiggling atoms is just equal to the energy the rock lost on impact.

Although the total energy is conserved when the rock stops, the energy *available for use* decreases. The kinetic energy of falling rocks, or falling water, could, for example, be used to turn a wheel. But once energy goes over to the random motion of atoms, it is unavailable to us except possibly as thermal energy. In any physical process, some energy becomes unavailable. When we're enjoined for environmental reasons to "conserve energy," we're being asked to use less *available* energy.

There is only one kind of kinetic energy, but there are many kinds of potential energy. The energy of that rock held at some height is gravitational potential energy. A compressed spring or a stretched rubber band has elastic potential energy. The elastic energy of the spring can, for example, be converted to kinetic energy in projecting a rock upward.

When positive and negative electrically charged objects are held apart from each other, those objects have electrical potential energy. If released, they would fly toward each other with increasing speed and kinetic energy. Planets orbiting the sun, or electrons orbiting the nucleus, have both kinetic and potential energy.

The chemical energy of a bottle of hydrogen and oxygen molecules is greater than the energy those molecules would have if they were bound together as water at the same temperature. Should a spark ignite that hydrogen–oxygen mixture, that greater energy would appear as kinetic energy of the resulting water molecules. The water vapor would therefore be hot. The chemical energy stored in the hydrogen–oxygen mixture would have become thermal energy.

Nuclear energy is analogous to chemical energy, except that the forces involved between the protons and the neutrons that make up the nucleus include nuclear forces as well as electrical forces. A uranium nucleus has a greater potential energy than do the fission products it breaks into. That greater potential energy becomes the kinetic energy of the fission products. That kinetic energy is thermal energy and can be used to make steam to turn turbines that turn generators to produce electric power. The potential energy of uranium can also be released rapidly as a bomb.

When light is emitted from a glowing hot body, energy goes into the electromagnetic radiation field, and the glowing body cools, unless it is supplied with additional energy. When a single atom emits light, it goes to a state of lower energy.

How many forms of energy are there? That depends on how you count. Chemical energy is, for example, ultimately electrical energy, though it is usually convenient to classify it separately. There may be forms of energy we don't yet know about. Several years ago it was discovered that the expansion of the universe is not slowing down, as had been generally believed. It's accelerating. The energy causing this acceleration has a name, "dark energy," but there is still more mystery about it than understanding.

What about "psychic energy"? Physics can claim no patent on the word "energy." It was used long before being introduced into physics in the early nineteenth century. If "psychic energy" could be converted into an energy treated by physics, it would be a form of the energy we're talking about. There is, of course, no generally accepted evidence for that.

Relativity

> Alice laughed. "There's no use trying," she said: "one *can't* believe impossible things."
>
> "I daresay you haven't had much practice," said the Queen. "When I was your age, I always did it for half-an-hour a day. Why sometimes I've believed as many as six impossible things before breakfast."
>
> —Lewis Carroll, *Through the Looking Glass*

When light became accepted as being a wave, it was assumed that something had to be waving. Electric and magnetic fields would then be distortions in this waving medium. Since material bodies moved through it without resistance, it was ethereal and was called the "ether." Since we receive light from the stars, the ether presumably pervaded the universe. Motion with respect to this ether would define an *absolute* velocity, something not meaningful without an ether to define a stationary "hitching post" in the universe.

In the 1890s Albert Michelson and Edward Morley set out to determine how fast our planet was moving through the universal ether. A boat moving in the same direction as the water waves sees the waves pass more slowly than when the boat moves in the direction opposite to the waves. From the difference in these two wave speeds, one can determine how fast the boat is moving on the water. This is essentially the experiment Michelson and Morley did with light waves.

To their surprise, Earth seemed not to be moving at all. At least, they measured the speed of light to be the same in all directions. Ingenious attempts to untangle this result using electromagnetic theory failed.

Albert Einstein took a different tack and cut the Gordian knot. He boldly *postulated* the observed fact: The speed of light is the same no matter how fast the observer moves. He took this strange result as a new property of Nature. Two observers, though moving at different speeds, would each measure the same light beam to be passing them at the same speed. The speed of light (in a vacuum) is therefore a universal constant, called "c."

With the speed of a light beam being the same for all observers, an absolute velocity could not be measured. Any observer, whatever his constant velocity, could consider himself to be at rest. There is then no absolute velocity; only *relative* velocities are meaningful. We therefore call it Einstein's "theory of relativity."

With just simple algebra, Einstein deduced further testable predictions from his postulate. The prediction most important to us in this book is that no object, no signal, no information, can travel faster than the speed of light. Another prediction is that mass is a form of energy that can be converted into other forms of energy. It's summarized as $E = mc^2$. Both of these predictions have been confirmed, sometimes dramatically.

The prediction of the theory of relativity that is hardest to believe is that the passage of time is relative: We see time passing more slowly for a fast-moving object than for one at rest.

Suppose a twenty-year-old woman travels to a distant star in her superfast rocketship, leaving her twin brother on Earth for thirty years. On her return, her brother, having aged thirty years, is now a middle-aged fifty. She, for whom time passed more slowly at her speed of, say, ninety-five percent that of light, has aged only ten years. She would be a relatively young thirty. The returned traveler would be twenty years younger than her stay-at-home twin in every physical and biological sense.

This "twin paradox" was raised early on as a supposed refutation of Einstein's theory. Could she not have considered *herself* at rest and her brother to have taken the speedy trip? He would then be younger than she. The theory, it was claimed, was inconsistent. Not so. The situation is not symmetric. Only observers moving at *constant* velocity (constant speed in a constant direction) can consider themselves at rest. That could not be true for the traveler, who had to turn around, to change her direction of motion, to accelerate, at the distant star in order to return home. (By the forces on her when she accelerated she could tell that she was not at rest.)

While it is not technically feasible to build rocketships to move people at near light speeds, relativity theory has been extensively tested and confirmed. Most tests have been with subatomic particles. The theory has also been checked by comparing accurate clocks flown around the world with clocks that stayed home. On their return, the traveling clocks were "younger." They recorded a bit less time by precisely the predicted amount. The validity of relativity theory is so well established today that only an extremely challenging test would be warranted. If you read about a test of "relativity," it is likely a test of the theory of *general* relativity, Einstein's theory of gravity. The full name of the theory we're talking about here is the theory of *special* relativity.

It's hard to believe the strange things that Einstein's relativity theory tells. That, for example, one could, in principle, become older than one's mother. But accepting the fact, now firmly established by experiment, that moving systems age less is good practice for believing the far stranger things that quantum mechanics tells.

We're now ready to start talking about those strange things.

Hello Quantum Mechanics

The universe begins to look more like a great thought than a great machine.

—Sir James Jeans

At the end of the nineteenth century, the search for Nature's basic laws seemed close to its goal. There was a sense of a task accomplished. Physics presented an orderly scene that fit the proper Victorian mood of the day.

Objects both on Earth and in the heavens behaved in accord with Newton's laws. So, presumably, did atoms. The nature of atoms was unclear. But to most scientists the rest of the job of describing the universe seemed a filling in of the details of the Great Machine.

Did the determinism of Newtonian physics deny "free will"? Physicists would leave such questions to philosophers. Defining the territory that physicists considered their own seemed straightforward. There was little to motivate a search for deeper meaning behind Nature's laws. But this intuitively reasonable worldview could not account for some puzzles physicists eventually saw in their laboratories. At first, the puzzles seemed merely details to be explored and resolved. Soon, however, the exploration challenged that comfortable classical view of the world. But today, a century later, the worldview is still in contention.

Quantum physics does not *replace* classical physics the way the sun-centered solar system replaced the earlier view with Earth as the cosmic center. Rather, quantum physics *encompasses* classical physics as a special case. Classical physics is usually an extremely good approximation for behavior of objects that are much larger than atoms. But if you dig deeply into any natural phenomenon, be it physical, chemical, biological, or cosmological, you hit quantum mechanics. The fundamental theories of physics from string theory to the Big Bang all start with quantum theory.

Quantum theory has been subject to challenging tests for eight decades. No prediction by the theory has ever been shown wrong. It is the most battle-tested theory in all of science. It has no competitors. Nevertheless, if you take the implications of the theory seriously, you confront an enigma. The theory tells us that the reality of the physical world depends somehow on our *observation* of it. This is hard to believe.

Being hard to believe presents a problem. If you're told something you can't believe, a likely response is: "I don't understand." In this case you may actually understand more than you think you do. We confront an enigma.

There is also a tendency to reinterpret what is said to make it seem reasonable. Don't use reasonableness as a test of comprehension. But here's one test: Niels Bohr, a founder of quantum theory, warned that unless you're shocked by quantum mechanics, you have not understood it.

Though our presentation may be novel, the experimental facts we describe and our quantum-theory explanations of those facts are completely undisputed. We step beyond that firm ground when we explore the *interpretation* of the theory and thus physics' encounter with consciousness. The deeper meaning of quantum mechanics is *increasingly* in dispute.

It does not require a technical background to move to the frontier where physics joins issues that seem beyond physics, and where physicists cannot claim unique competence. Once there, you can take sides in the debate.

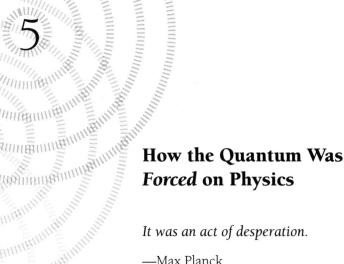

How the Quantum Was *Forced* on Physics

It was an act of desperation.

—Max Planck

Physics courses are rarely presented historically. The introductory course in quantum mechanics is the exception. For students to see why we accept a theory so violently in conflict with common sense, they must see how physicists were dragged from their nineteenth-century complacency by the brute facts observed in their laboratories.

The Reluctant Revolutionary

In the final week of the nineteenth century, Max Planck suggested something outrageous: The most fundamental laws of physics were violated. This was the first hint of the quantum revolution, that the worldview we now call "classical" had to be abandoned.

Max Planck, son of a distinguished professor of law, was careful, proper, and reserved. His clothes were dark and his shirts stiffly starched. Raised in the strict Prussian tradition, Planck respected authority, both in society and in science. Not only should people rigorously obey the laws, so should physical matter. Not your typical revolutionary.

In 1875, when young Max Planck announced his interest in physics, the chairman of his physics department suggested he study something more exciting. Physics, he said, was just about complete: "All the important discoveries have already been made." Undeterred, Planck completed his studies in physics and plugged away for years as a *Privatdozent*, an

Figure 5.1 Max Planck

apprentice professor, receiving only the small fees paid by students attending his lectures.

Planck chose to work in the most properly lawful area of physics, thermodynamics, the study of heat and its interaction with other forms of energy. His solid but unspectacular work eventually won him a professorship. His father's influence supposedly helped.

A nagging unexplained phenomenon in thermodynamics was thermal radiation: the spectrum, the colors of the light given off by hot bodies. The problem was one of Kelvin's "two dark clouds on the horizon." Planck set about to solve it.

We'll first look at some aspects that seemed reasonable, and then the problem. That a hot poker should glow seems obvious. At the turn of the century, although the nature of atoms, even the existence of atoms, was unclear, electrons had just been discovered. Presumably these little charged particles jiggled in a hot body and therefore emitted electromagnetic radiation. Because this radiation was the same no matter what material it came from, it seemed a fundamental aspect of Nature and therefore important to understand.

It seemed reasonable that, as a piece of iron got hotter, its electrons should shake harder and, presumably, at a higher rate, meaning at a higher frequency. Therefore, the hotter the metal, the brighter it glows, and glows at a higher frequency. As the iron gets hotter, its color thus goes from the invisible infrared, to a visible red, then to orange, and eventually the metal becomes white hot as the emitted light covers the entire visible frequency range.

Since our eyes can't see frequencies above the violet, superhot objects, which emit mostly in the ultraviolet, appear bluish. Materials on Earth vaporize before they get hot enough to glow blue, but we can look up and see hot blue stars. Even cool objects "glow," though weakly and at low frequencies. Bring your palm close to your cheek and feel the warmth from the infrared light your hand emits. The sky shines down on us with invisible microwave radiation left over from the flash of the Big Bang.

In figure 5.2, we sketch the actual intensity of radiation from the sun's 6,000°C surface at different frequencies, which we just label as colors. A hotter object emits more light at all frequencies, its maximum intensity

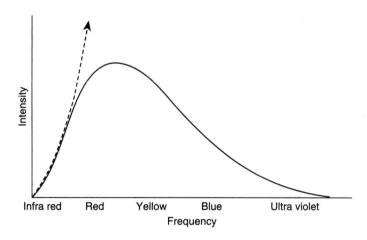

Figure 5.2 6,000°C thermal radiation (solid line) compared with
the classical prediction (dashed line)

is at a higher frequency, but the intensity always drops at very high frequencies.

The dashed line shows the problem. It is the theoretical intensity calculated with the laws of physics accepted in 1900. Notice that the theory and the experimental observations agree in the infrared. But at higher frequencies, the classical physics calculation not only gave a wrong answer, it gave a *ridiculous* answer. It predicted a forever increasing light intensity at frequencies beyond the ultraviolet.

Were this true, every object would instantaneously lose its heat by radiating a burst of energy at frequencies beyond the ultraviolet. This embarrassing deduction was derided as the "ultraviolet catastrophe." But no one could say where the seemingly sound reasoning leading to it went wrong.

Planck struggled for years to derive a formula from classical physics that fit the experimental data. In frustration, he decided to attack the problem backwards. He would first try to *guess* a formula that agreed with the data and then, with that as a hint, try to develop a proper theory. In a single evening, studying the data others had given him, he found a quite simple formula that worked perfectly.

If Planck put in the temperature of the body, his formula gave the correct radiation intensity at every frequency. His formula needed a "fudge factor" to make it fit the data, a number he called "h." We now call it "Planck's constant" and recognize it as a fundamental property of Nature, like the speed of light.

With his formula as a hint, Planck sought to explain thermal radiation in terms of the basic principles of physics. According to the straightforward ideas of the day, an electron would start vibrating if it were bumped by a jiggling neighboring atom in a hot metal. This little charged particle would then gradually lose its energy by emitting light. We plot such an energy loss in figure 5.3. In a similar fashion, a pendulum bob on a string, or a child on a swing, given a shove, would continuously lose energy to air resistance and friction.

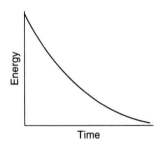

Figure 5.3 Energy loss by charged particle according to classical physics

However, every description of electrons radiating energy according to the physics of the day led to that same crazy prediction, the ultraviolet catastrophe. After a long struggle, Planck ventured an assumption that violated the universally accepted principles of physics. At first, he didn't take it seriously. He later called it "an act of desperation."

Max Planck assumed an electron could radiate energy only in chunks, in "quanta" (the plural of quantum). Each quantum would have an energy equal to the number *h* in his formula times the vibration frequency of the electron.

Behaving this way, an electron would vibrate for a while without losing energy to radiation. Then, randomly, and *without cause*, without an impressed force, it would suddenly radiate a quantum of energy as a pulse of light. (Electrons would also gain their energy from the hot atoms by such "quantum jumps.") In figure 5.4 we plot an example of such energy loss in sudden steps. The dashed line repeats the classically predicted, gradual energy loss of figure 5.3.

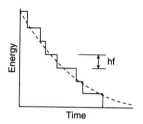

Figure 5.4 Energy loss by charged particle according to Planck

Planck was allowing electrons to violate both the laws of electromagnetism and Newton's universal equation of motion. Only by this wild assumption could he derive the formula he had guessed, the formula that correctly described thermal radiation.

If this quantum-jumping behavior is indeed a law of Nature, it should apply to everything. Why, then, do we see the things around us behaving smoothly? Why don't we see children on swings suddenly change their swinging motion in quantum jumps? It's a question of numbers, and *h* is an extremely small number.

Not only is *h* small, but since the frequency of a child moving back and forth on a swing is much lower than the frequency at which an electron vibrates, the quantum steps of energy (*h* times frequency) are vastly smaller for the child. And, of course, the total energy of a swinging child is vastly larger than that of an electron. Therefore, the number of quanta involved in the child's motion is vastly, vastly greater than the number of quanta involved in the motion of the electron. A quantum jump, the change in energy by a single quantum, is thus far too small to be seen for the child on a swing.

But let's go back to Planck's day and the reaction to the solution he proposed for the thermal radiation problem. His formula fit the experimental data well. But his explanation seemed more confounding than the problem it presumed to solve. Planck's theory seemed silly. But no one laughed, at least not in public—Herr Professor Planck was too important a man for that. His quantum-jumping suggestion was simply ignored.

Physicists were not about to challenge the fundamental laws of mechanics and electromagnetism. Even if the classical laws gave a ridiculous prediction for the light emitted by glowing bodies, these basic principles seemed to work everyplace else. And they made sense. Planck's colleagues felt a reasonable solution would eventually be found. Planck himself agreed and promised to seek one. The quantum revolution arrived with an apology, and almost unnoticed.

In later years, Planck even came to fear the negative *social* consequences of quantum mechanics. Freeing the fundamental constituents of matter from the rules of proper behavior might seem to free people from responsibility and duty. The reluctant revolutionary would have liked to cancel the revolution he sparked.

The Technical Expert, Third Class

His parents worried about mental retardation when young Albert Einstein was slow in starting to talk. Later, though, he became an avid and independent student of things that interested him. But his distaste for the rote instruction at the *Gymnasium* (high school) led to his not doing well. Asked to suggest a profession that Albert might follow, the headmaster confidently predicted: "It doesn't matter; he'll never make a success of anything."

Einstein's parents left Germany for Italy after the family electrochemical business failed. The new business in Italy fared little better. Young Einstein was soon on his own. He took the entrance exam to the Zurich Polytechnical Institute but did not pass. He was finally admitted the next year. On graduation, he was unsuccessful in trying for a position as *Privatdozent*. He had the same luck in applying for a teaching job at the *Gymnasium*. For a while Einstein supported himself as a tutor for students having trouble with high school. Eventually, through a friend's influence, he got a job in the Swiss patent office.

His duties as Technical Expert, Third Class, were to write summaries of patent applications for his superiors to use in deciding whether an idea

Figure 5.5 Albert Einstein. Courtesy
California Institute of Technology and
the Hebrew University of Jerusalem

warranted a patent. Einstein enjoyed the work, which did not take his full time. Keeping an eye on the door in case a supervisor came in, he also worked on his own projects.

Initially, Einstein continued on the subject of his doctoral thesis, the statistics of atoms bouncing around in a liquid. This work soon became the best evidence for the atomic nature of matter, something still debated at the time. Einstein was struck by a mathematical similarity between the equation for the motion of atoms and Planck's radiation law. He wondered: Might light not only be mathematically like atoms, but also be physically like atoms?

If so, might light, like matter, come in compact lumps? Perhaps the pulses of light energy emitted in one of Planck's quantum jumps did not expand in all directions as Planck assumed. Could the energy instead be confined to a small region? Might there be atoms of light as well as atoms of matter?

Einstein speculated that light is a stream of compact lumps, "photons" (a term that came later). Each photon would have an energy equal to Planck's quantum hf (Planck's constant, h, times the light's frequency). Photons would be created when electrons emit light. Photons would disappear when light is absorbed.

Seeking evidence that his speculation might be right, Einstein looked for something that might display a granular aspect to light. It was not hard to find. The "photoelectric effect" had been known for almost twenty years. Light shining on a metal could cause electrons to pop out.

The situation was messy. Unlike thermal radiation, where a universal rule held for all materials, the photoelectric effect was different for each substance. Moreover, the data were inaccurate and not particularly reproducible.

Never mind the bad data. Spread-out light *waves* shouldn't kick electrons out of a metal at all. Electrons are too tightly bound. While electrons are free to move about within a metal, they can't readily escape it. We can "boil" electrons out of a metal, but it takes a very high temperature. We can pull electrons out of a metal, but it takes a very large electric field. Nevertheless, dim light, corresponding to an extremely weak electric field, still ejects electrons. The dimmer the light, the fewer the electrons. But no matter how dim the light, some electrons were always ejected.

Einstein gleaned even more information from the bad data. Electrons popped out with high energy when the light was ultraviolet or blue. With lower frequency yellow light, their energy was less. Red light usually ejected no electrons. The higher the frequency of the light, the greater the energy of the emitted electrons.

The photoelectric effect was just what Einstein needed. Planck's radiation law implied that light was emitted in pulses, quanta, whose energy was larger for higher frequency light. If the quanta were actually compact lumps, all the energy of each photon might be concentrated on a single electron. A single electron absorbing a whole photon would gain a whole quantum of energy hf.

Light, especially high-frequency light with its high-energy photons, could then give electrons enough energy to jump out of the metal.

The higher the energy of the photon, the higher the energy of the ejected electron. For light below a certain frequency, its photons would have insufficient energy to remove an electron from the metal, and no electrons would be ejected.

Einstein said it clearly in 1905:

> According to the presently proposed assumption the energy in a beam of light emanating from a point source is not distributed continuously over larger and larger volumes of space but consists of a finite number of energy quanta, localized at points of space which move without subdividing and which are absorbed and emitted only as units.

Assuming that light comes as a stream of photons and that a single electron absorbs all the energy of a photon, Einstein used the conservation of energy to derive a simple formula relating the frequency of the light to the energy of the ejected electrons. We plot it in figure 5.6. Photons with energy less than the energy binding the electrons into the material could not kick any electrons out at all.

A striking aspect of Einstein's photon hypothesis is that the slope of the straight line on this graph is just Planck's constant, h. Until this time, Planck's constant was just a number needed to fit Planck's formula to the observed thermal radiation. It appeared nowhere else in physics. Before Einstein's photon hypothesis, there was no reason to think the ejection of electrons by light had anything at all to do with the radiation emitted by hot bodies. This slope was the first indication that the quantum was universal.

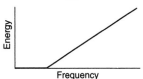

Figure 5.6 Energy of ejected electrons versus light frequency

Ten years after Einstein's work on the photoelectric effect, the American physicist Robert Millikan found that Einstein's formula in every case predicted "exactly the observed results." Nevertheless, Millikan called Einstein's photon hypothesis leading to that formula "wholly untenable" and called Einstein's suggestion that light came as compact particles "reckless."

Millikan was not alone. The physics community received the photon postulate "with disbelief and skepticism bordering on derision." However, eight years after proposing the photon, Einstein had gained a considerable reputation as a theoretical physicist for many other achievements and was nominated for membership in the Prussian Academy of Science. Nevertheless, Planck, in his letter supporting that nomination, felt he had to defend Einstein: "[T]hat he may sometimes have missed the target in his speculations, as, for example, in his hypothesis of light quanta, cannot really be held too much against him. . . ."

Even when Einstein was awarded the Nobel Prize in 1922 for the photoelectric effect, the citation avoided explicit mention of the then seventeen-year-old, but still unaccepted, photon. An Einstein biographer writes: "From 1905 to 1923, [Einstein] was a man apart in being the only one, or almost the only one, to take the light-quantum seriously." (We tell what happened in 1923 later in this chapter.)

Though the reaction of the physics community to Einstein's photons was, in a word, rejection, they were not just pig-headed. Light was *proven* to be a spread-out wave. Light displayed interference. A stream of discrete particles could not do that.

Recall our discussion of interference in chapter 4: Light coming through a single narrow slit illuminates a screen more or less uniformly. Open a second slit, and a pattern of light and dark bands appears whose spacing depends on the separation of the two slits. At those dark places, wave crests from one slit arrive together with wave troughs from the other.

Waves from one slit thus cancel waves from the other. Interference demonstrates that light is a wave spread out over both slits.

In chapter 4 we mentioned that the argument that tiny bullets could not cause interference was not airtight. Might they not somehow *deflect* each other to form the bright and dark bands? That loophole in the argument has been closed. Now that we know how much energy each photon carries, we can know how many photons are in a beam of a given intensity. We see interference with light so dim, so low intensity, that only one photon is present in the apparatus at a time.

Figure 5.7 Interference pattern formed by light coming through two narrow slits

Choosing to demonstrate interference, something explicable *only* in terms of waves, you could demonstrate light to be a widely spread-out wave.

However, by choosing a photoelectric experiment, you could demonstrate the opposite: that light was *not* a spread-out wave, but rather a stream of tiny compact objects. There seems to be an inconsistency. (Recall something like this in Neg Ahne Poc: Our visitor could choose to demonstrate that the couple was spread over both huts, one person in each, or he could choose to demonstrate that the couple was compactly concentrated in a single hut.)

Though the paradoxical nature of light disturbed Einstein, he clung to his photon hypothesis. He declared that a mystery existed in Nature and that we must confront it. He did not pretend to resolve the problem. And we do not pretend to resolve it in this book. The mystery is still with us one hundred years later. Later chapters focus on the implication of our being able to choose to establish either of two contradictory things. The mystery extends beyond physics to the nature of observation. It's the quantum enigma. Far-out speculations are seriously proposed today by distinguished experts in quantum physics.

In a single year, 1905, Einstein discovered the quantum nature of light, firmly established the atomic nature of matter, and formulated the theory of relativity. The following year the Swiss patent office promoted Einstein: to Technical Expert, *Second* Class.

The Postdoc

Niels Bohr grew up in a comfortable and respected family that nurtured independent thought. His father, an eminent professor of physiology at Copenhagen University, was interested in philosophy and science and encouraged those interests in his two sons. Niels's brother, Harald, eventually became an outstanding mathematician. Niels Bohr's early years were supportive. Unlike Einstein, Bohr was never the rebel.

In college in Denmark, Bohr won a medal for some clever experiments with fluids. But we skip ahead to 1912 when, with his new Ph.D., Bohr went to England as a "postdoc," a postdoctoral student.

By this time the atomic nature of matter was generally accepted, but the atom's internal structure was unknown. Actually, it was in dispute. Electrons, negatively charged particles thousands of time lighter than any atom, had been discovered a decade earlier by J. J. Thompson. An atom,

Figure 5.8 Niels Bohr. Courtesy the American Institute of Physics

being electrically neutral, must somewhere have a positive charge equal to that of its negative electrons, and that positive charge presumably had most of the mass of the atom. How were the atom's electrons and its positive charge distributed?

Thompson made the simplest assumption: The massive positive charge uniformly filled the atomic volume, and the electrons, one in hydrogen and almost 100 in the heaviest known atoms, were supposedly distributed throughout the positive background, like raisins in a rice pudding. Theorists tried to calculate how various distributions of electrons might give each element its characteristic properties.

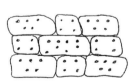

Figure 5.9 Thompson's rice pudding model of atoms

There was a competing model for the atom. Ernest Rutherford at the University of Manchester in England explored the atom by shooting alpha particles (helium atoms stripped of their electrons) through a gold foil. He saw something inconsistent with Thompson's uniformly distributed positive mass. About one alpha in 10,000 would bounce off at a large angle, sometimes even backward. The experiment was likened to shooting prunes through rice pudding. Collisions with the small raisins (electrons) could not knock a fast prune (an alpha) much off track. Rutherford concluded that his alpha particles were colliding with an atom's massive positive charge, which was concentrated in a small lump at the center of the atom, a "nucleus."

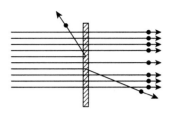

Figure 5.10 Rutherford's experiment with alpha particles

Why did the negative electrons, attracted by the positive nucleus, not just crash down into it? Presumably, for the same reason that planets don't crash down into the sun: Planets *orbit* the sun. Rutherford decided that electrons orbited a compact, massive, positive nucleus.

There was a problem with Rutherford's planetary model: instability. Since an electron is charged, it should radiate as it races around its orbit. Calculations showed that an electron should give off its energy as light and spiral down to crash into the nucleus in less than a millionth of a second.

Figure 5.11 Instability of Rutherford's atomic model

Most of the physics community considered the instability in the planetary model a more serious problem than the rice pudding model's inability to explain the rare large-angle deflections of Rutherford's alpha particles. But Rutherford, a supremely confident fellow, *knew* his planetary model was basically right.

When the young postdoc Bohr arrived in Manchester, Rutherford assigned him the job of explaining how the planetary atom might be stable. Bohr's tenure in Manchester lasted only six months, supposedly because his support money ran out. But an eagerness to get back to Denmark to marry the beautiful Margrethe likely shortened his stay. While teaching at the University of Copenhagen in 1913, Bohr continued to work on the stability problem.

How he got his successful idea is not clear. But while other physicists were trying to understand how the quantum of energy and Planck's constant, h, arose from the classical laws of physics, Bohr took an "h okay!" attitude. He just accepted quantization as fundamental. After all, it had worked for Planck, and it had worked for Einstein.

Bohr wrote a very simple formula that said that "angular momentum," the rotational motion of an object, could exist only in quantum units. If so, only certain electron orbits were allowed. And, most important, he wrote his formula so that there was a smallest possible orbit. By fiat, Bohr's formula "forbids" an electron to crash into the nucleus. If his ad hoc formula was correct, the planetary atom was stable.

Without more evidence, Bohr's quantum idea would be rejected out of hand. But with his formula Bohr could readily calculate all the energies allowed for the single electron orbiting a proton, the nucleus of the hydrogen atom. From those energies he could then calculate the particular frequencies, or colors, of light that could be emitted from hydrogen atoms electrically excited in a "discharge," something like a neon sign only with hydrogen gas inside instead of neon.

Those frequencies had been carefully studied for years, though Bohr was initially unaware of that work. Why only certain frequencies were emitted was a complete mystery. The spectrum of frequencies, unique to each element, presented a pretty set of colors. But were they any more significant than the particular patterns of butterfly wings? Now, however, Bohr's quantum rule predicted the frequencies for hydrogen with stunning accuracy: precise to parts in 10,000. Although at this time Bohr's theory had light emitted by atoms in energy quanta, he, along with essentially all other physicists, still rejected Einstein's compact photon.

Some physicists dismissed Bohr's theory as "number juggling." Einstein, however, called it "one of the greatest discoveries." And others soon came to agree. Bohr's basic idea was rapidly applied widely in physics and chemistry. No one understood *why* it worked. But work it did. And for Bohr that was the important thing. Bohr's pragmatic "h okay!" attitude toward the quantum quickly brought him success.

Contrast Bohr's early triumph with his quantum ideas with Einstein's long remaining "a man apart" in his belief in the almost universally rejected photon. Notice, in later chapters, how the early experiences of these two men is reflected in their lifelong friendly debate about quantum mechanics.

The Prince

Louis de Broglie was *Prince* Louis de Broglie. His aristocratic family intended a career in the French diplomatic service for him, and young Prince Louis studied history at the Sorbonne. But after receiving an arts degree, he moved to theoretical physics. Before he could do much physics, World War I broke out, and de Broglie served in the French army at a telegraph station in the Eiffel Tower.

With the war over, de Broglie started work on his physics Ph.D., attracted, he says, "by the strange concept of the quantum." Three years into his studies, he read the recent work of the American physicist Arthur Compton. An idea clicked in his head. It led to a short doctoral thesis, and eventually to a Nobel Prize.

Figure 5.12 Louis de Broglie. Courtesy
the American Institute of Physics

Compton had, in 1923, almost two decades after Einstein proposed the photon, discovered, to his surprise, that when light bounced off electrons its frequency changed. This is not wave behavior: When a wave reflects from a stationary object, each incident crest produces one other wave crest. The frequency of the wave therefore does not change in reflection. On the other hand, if Compton assumed that light was a stream of particles, *each with the energy of an Einstein photon*, he got a perfect fit to his data.

The "Compton effect" did it! Physicists now accepted photons. Sure, in certain experiments light displayed its spread-out wave properties and in others its compact particle properties. As long as one knew under what conditions each property would be seen, the photon idea seemed less troublesome than finding another explanation for the Compton effect. Einstein, however, remained "a man apart." He insisted a mystery remained, once saying: "Every Tom, Dick, and Harry thinks they know what the photon is, but they're wrong."

Graduate student de Broglie shared Einstein's feeling that there was a deep meaning to light's duality, being *either* a spread-out wave or a stream of compact particles. He wondered whether there might be symmetry in Nature. If light was either wave or particle, perhaps matter was also either particle or wave. He wrote a simple expression for the wavelength of a particle of matter. This formula for the "de Broglie wavelength" of a particle is something every beginning quantum mechanics student quickly learns.

	Wave	Particle
Light	✓	✓
Matter	?	✓

Figure 5.13 De Broglie's symmetry idea

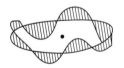

Figure 5.14
Wavelengths around an electron orbit

The first test of that formula came from a puzzle that stimulated de Broglie's wave idea: If an electron in a hydrogen atom were a compact particle, how could it possibly "know" the size of the orbit it should move in to exist in only those orbits allowed by Bohr's by-then-famous formula?

The lengths of violin string required to produce a given pitch are determined by the number of half-wavelengths of the vibration that fit along the length of the string. Similarly, with the electron as a wave, the allowed orbits might be determined by the number of electron wavelengths that fit around the orbit's circumference. Applying this idea, de Broglie was able to *derive* Bohr's previously ad hoc quantum rule. (In the violin, it's the material of the string that vibrates. What vibrates in the case of the electron "wave" was then a mystery. It still is.)

It's not clear how seriously de Broglie took his conjecture. He certainly did not recognize it as advancing a revolutionary view of the world. In his own later words:

> [H]e who puts forward the fundamental ideas of a new doctrine often fails to realize at the outset all the consequences; guided by his personal intuitions, constrained by the internal force of mathematical analogies, he is carried away, almost in spite of himself, into a path of whose final destination he himself is ignorant.

De Broglie took his speculation to his thesis adviser, Paul Langevin, famous for his work on magnetism. Langevin was not impressed. He noted that in deriving Bohr's formula de Broglie merely replaced one ad hoc assumption with another. Moreover, de Broglie's assumption, that electrons could be waves, seemed ridiculous.

Were de Broglie an ordinary graduate student, Langevin might have summarily dismissed his idea. But he was *Prince* Louis de Broglie. Aristocracy was meaningful, even in the French republic. So no doubt to cover himself, Langevin asked for a comment on de Broglie's idea from the world's most eminent physicist. Einstein replied that this young man has "lifted a corner of the veil that shrouds the Old One."

Meanwhile, there was a minor accident in the laboratories of the telephone company in New York. Clinton Davisson was experimenting with the scattering of electrons from metal surfaces. While Davisson's interests were largely scientific, the phone company was developing vacuum tube amplifiers for telephone transmissions, and for that, the behavior of electrons striking metal was important.

Electrons usually bounced off the normally amorphous metal surface in all directions. But after the accident, in which air leaked into his vacuum system and oxidized a nickel surface, Davisson heated the metal to drive off the oxygen. The nickel crystallized, essentially forming an array of slits. Electrons now bounced off in only a few well-defined directions. It was an interference pattern demonstrating the electron's wave nature. The discovery confirmed de Broglie's speculation that material objects could also be waves.

We opened this chapter with the first hint of the quantum in 1900. It was a hint largely ignored. We close it with physicists in 1923 finally forced to accept a wave–particle duality: A photon, an electron, an atom, a molecule, in principle any object, can be either compact or widely spread out. You can show on object to be either bigger than a loaf of bread or smaller than an atom. You can choose which of these two contradictory features to demonstrate. The physical reality of an object depends on how you *choose* to look at it.

Physics had encountered consciousness but did not yet realize it. Awareness of that contact came a few years later, after Erwin Schrödinger's discovery of the new universal law of motion. That discovery is the subject of our next chapter.

6

Schrödinger's Equation
The *New* Universal Law of Motion

If we are still going to put up with these damn quantum jumps, I am sorry that I ever had anything to do with quantum theory.

—Erwin Schrödinger

By the early 1920s, physicists had accepted that electrons, and presumably other matter as well as light, could be demonstrated to be *either* compact lumps *or* widely spread-out waves. It depended on the experiment you chose to perform.

Since Einstein's 1905 photon explanation of the photoelectric effect, the undisputed experimental facts were right there in front of physicists. But the implications of those facts were largely ignored. In 1909 Einstein emphasized that the light quantum posed a serious problem. But as "a man apart," he was almost the only one to take the light quantum seriously. In 1913, Bohr talked of light being emitted in quantum jumps, but he did not accept the compact photon. In 1915 Milliken had called Einstein's photon proposal "reckless." However, with Compton's 1923 scattering of individual photons by electrons, physicists quickly accepted the photon. They nevertheless ignored Einstein's persistent concern. Why? No doubt they expected that a fundamental theory, a theory still to come, would resolve the troublesome "wave–particle duality" paradox. The fundamental theory soon came, but it brought no resolution—quite the opposite.

Recognition of the paradox as a serious problem came three years later, in 1926, with the Schrödinger equation. Erwin Schrödinger was not looking to resolve the wave-particle paradox. He saw de Broglie's matter waves as a way to get rid of Bohr's "damn quantum jumps." He would explain matter waves.

Erwin Schrödinger, the only child of a prosperous Viennese family, was an outstanding student. As an adolescent his interest was in theater and art. Rebelling against the bourgeois society of late-nineteenth-century Vienna, Schrödinger rejected the Victorian morality of his upbringing. Throughout his life he pursued intense romances, his lifelong marriage notwithstanding.

After serving in the First World War as a lieutenant in the Austrian army on the Italian front, Schrödinger started teaching at the University of Vienna. About this time he embraced the Indian mystical teaching of Vedanta, but he seems to have kept this philosophical leaning separate from his physics. In 1927, just after his spectacular work in quantum mechanics, he was invited to Berlin University as Planck's successor. With Hitler's coming to power in 1933, Schrödinger, though not Jewish, left Germany.

Figure 6.1 Erwin Schrödinger. Courtesy the American Institute of Physics

After visits to England and the United States, he incautiously returned to Austria to accept a position at the University of Graz. With Hitler's annexation of Austria, he was in trouble. Leaving Germany had established his opposition to the Nazis. Escaping through Italy, he spent the rest of his career at the School for Theoretical Physics in Dublin, Ireland.

In his middle years, Schrödinger's thoughts turned to include questions of what quantum mechanics implied *beyond* physics. He produced two short but extremely influential books. In *What Is Life?* he suggested quantum mechanical reasons for the source of genetic inheritance being an "aperiodic crystal." Francis Crick, co-discoverer of the structure of DNA, credits Schrödinger's book for inspiration. The first chapter of Schrödinger's other book, *Mind and Matter*, is titled "The Physical Basis of Consciousness."

A Wave Equation

Despite the successes of the early quantum theory based on Bohr's quantum rule, Schrödinger rejected a physics where electrons moved only in "allowed orbits" and then, without cause, abruptly jumped from one orbit to another. He was outspoken:

> You surely must understand, Bohr, that the whole idea of quantum jumps necessarily leads to nonsense. It is claimed that the electron in a stationary state of an atom first revolves periodically in some sort of an orbit without radiating. There is no explanation of why it should not radiate; according to Maxwell's theory, it must radiate. Then the electron jumps from this orbit to another one and thereby radiates. Does the transition occur gradually or suddenly? . . . And what laws determine its motion in a jump? Well, the whole idea of quantum jumps must simply be nonsense.

Schrödinger credits Einstein's "brief but infinitely far-seeing remarks" for calling his attention to de Broglie's speculation that material objects could display a wave nature. The idea appealed to him. Waves might evolve smoothly from one state to another. Electrons as waves would not need to orbit without radiating. He would get rid of Bohr's "damn quantum jumps."

Perfectly willing to amend Newton's laws to account for quantum behavior, Schrödinger nevertheless wanted a description of the world that had electrons and atoms behaving *reasonably*. He would seek an equation governing waves of matter. It would be new physics, therefore a guess that would have to be tested. Schrödinger would seek the *new* universal equation of motion.

A *universal* equation would have to work for large objects as well as small. From the position and motion of a tossed stone at one moment, Newton's law predicts the stone's future position and motion. Similarly, from a wave's initial shape, a wave equation predicts the wave's shape at any later time. It describes how ripples spread from the spot where a tossed pebble hits the water, or how waves propagate on a taut rope.

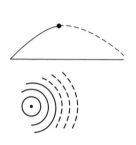

Figure 6.2 The path of a stone and the spreading of water ripples

There was a problem: The wave equation that works for waves in water, and also for waves of light and sound, doesn't work for matter waves. Waves of light and sound move at the single speed determined by the medium in which the wave propagates. Sound, for example, moves at 330 meters per second in air. The wave equation Schrödinger sought had to allow matter waves to move at *any* speed because electrons, atoms— and baseballs—move at any speed.

The breakthrough came during a mountain vacation with a girlfriend in 1925. His wife stayed home. To aid his concentration, Schrödinger brought with him two pearls to keep noise out of his ears. Exactly what noise he wished to avoid is not clear. Nor do we know the identity of the girlfriend, or whether she was inspiration or distraction. Schrödinger kept discreetly coded diaries, but the one for just this period is missing.

In four papers published within the next six months, Schrödinger laid down the basis of modern quantum mechanics with an equation describing waves of matter. The work was immediately recognized as a triumph. Einstein said it sprang from "true genius." Planck called it "epoch making." Schrödinger himself was delighted to think that he had gotten rid of quantum jumping. He wrote:

> It is hardly necessary to point out how much more gratifying it would be to conceive a quantum transition as an energy change

from one vibrational mode to another than to regard it as a jumping of electrons. The variation of vibrational modes may be treated as a process continuous in space and time and enduring as long as the emission process persists.

(The Schrödinger equation is actually a nonrelativistic approximation. That is, it holds only when speeds are not close to that of light. The conceptual issues we treat are still with us in the more general case. It is simpler, clearer, and also customary to deal with the quantum enigma in terms of the Schrödinger equation. And even though photons move at the speed of light, essentially everything we say applies to photons.)

History is more complicated than the story we just told, and more acrimonious. Almost simultaneously with Schrödinger's discovery, Bohr's young postdoc, Werner Heisenberg (of whom we'll hear more later), presented his own version of quantum mechanics. It was an abstract mathematical method for obtaining numerical results. It denied any pictorial description of what was going on. Schrödinger criticized Heisenberg's approach: "I was discouraged, if not repelled, by what appeared to me a rather difficult method of transcendental algebra, defying any visualization." Heisenberg was equally unimpressed by Schrödinger's wave picture. In a letter to a colleague: "The more I ponder the physical part of Schrödinger's theory, the more disgusting it appears to me."

For a while it seemed that two intrinsically different theories explained the same physical phenomena, a disturbing possibility that philosophers had long speculated about. But within a few months, Schrödinger proved that Heisenberg's theory was logically identical to his own, just a different mathematical representation. The more mathematically tractable Schrödinger version is generally used today.

The Wavefunction

Heisenberg did, however, have a point about the physical aspect of Schrödinger's theory. What's waving in Schrödinger's matter wave? The mathematical representation of the wave is called the "wavefunction." In some sense, the wavefunction of an object is the object itself. In standard quantum theory no atom exists in addition to the wavefunction of the atom.

But what, exactly, is Schrödinger's wavefunction *physically*? At first, Schrödinger didn't know, and when he speculated, he was wrong. For now, let's just plow ahead and look at some wavefunctions that the equation tells us can exist. We'll worry later about what they actually are physically. That's what Schrödinger did.

We first consider the wavefunction of a simple little object moving along in a straight line. It could be an electron or an atom, for example. To be general, we often refer to an "object" but sometimes we revert to "atom." We later discuss wavefunctions for bigger things—a molecule, a baseball, a cat, even the wavefunction of a friend. Cosmologists contemplate the wavefunction of the universe, and so will we.

A couple of years before Schrödinger's vacation inspiration, Compton had shown that photons bounced off electrons as if electrons and photons were like tiny, compact billiard balls. On the other hand, to display interference, each and every photon or electron had to be a widely spread-out wave coming on two paths. How can a single object be *both* compact *and* spread out? A wave can be *either* compact or spread out. But not both compact and spread out at the same time. What was an atom, an electron, or a photon really like? Was the atom a compact object or a spread-out object? There was still a problem.

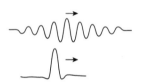

Figure 6.3 Wavefunction as a series of waves or a single crest

But one thing worked out nicely: For big things, objects much larger than atoms, Schrödinger's equation essentially *becomes* Newton's universal equation of motion. Schrödinger's equation thus governs not only the behavior of electrons and atoms but also the behavior of everything made of atoms—molecules, baseballs, and planets. The Schrödinger equation tells what the wavefunction will be in a given situation, and how it will change with time. It's the *new* universal law of motion. Newton's law of motion is just the excellent approximation for big things.

Waviness

Schrödinger's equation says a moving object is a moving wave packet. But, once more, what's waving? Think of these analogies—Schrödinger no doubt did:

At a stormy place in the ocean, the waves are big. We'll call that a region of large "waviness." The boom of a drum, on its way to you from a distant drummer, is where the air pressure waviness is large; it's where the sound *is*. The bright patch where the sunlight hits the wall, the region of large electric field waviness, is where the light *is*. In these cases, waviness tells where something *is*. It would seem reasonable to carry this notion over to the quantum case.

The waviness of a packet of quantum waves is large where the amplitude of the waves is large, where the crests are high and the troughs deep. Waviness can be easy to sketch if we have the wavefunction. We will indicate waviness by shading: The more the shading, the greater the waviness. (The mathematical term for the waviness is the "absolute square of the wavefunction," and there is a mathematical procedure for getting it from the wavefunction. We mention this only because you might see that term elsewhere. "Waviness" is more descriptive.)

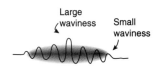

Figure 6.4 A wavefunction and its waviness

When we consider an atom simply as an object moving along in one direction, we ignore its internal structure. There are, however, electron wavefunctions *within* the atom. Early on, Schrödinger calculated the wavefunction of the single electron in the hydrogen atom. He duplicated Bohr's results for the energy levels and the experimentally observed hydrogen spectrum. Able to do that without needing Bohr's arbitrary assumptions, an elated Schrödinger was sure he had it right. He had gotten rid of quantum jumps, he thought. Not so, we'll see.

In figure 6.5, we sketch the waviness for the hydrogen electron's three lowest energy states as cross sections through the three-dimensional waviness of the electron. You can visualize the waviness as clumps of fog. The fog is densest where the waviness is largest. The shape of the fog clump is, in a sense, the shape of the atom. Calculated pictures such as these provide chemists with insight into how atoms and molecules bind with each other.

Few of us thought the waviness of electrons *within* atoms would ever be displayed

Figure 6.5 The waviness of a hydrogen atom's three lowest states

directly, the way an interference pattern displays the widely spread-out waviness of free electrons or atoms. The patterns of figure 6.5 are calculated from the Schrödinger equation and are then indirectly confirmed by the behaviors they infer. In 2009, Ukrainian physicists, using an old imaging technique, "field-emission microscopy," pulled electrons out of single carbon atoms with a large electric field. By noting where on a detection screen the electrons landed, they could trace back to the position *inside the atom* from which the electrons emerged. They directly confirmed familiar textbook patterns of waviness.

Have we suggested that the waviness tells where the spread-out object *is*? It's not quite that.

Schrödinger's Initial (Wrong) Interpretation of Waviness

Schrödinger speculated that an object's waviness was the smeared out object itself. Where, for example, the electron fog is densest, the material of the electron would be most concentrated. The electron itself would thus be smeared over the extent of its waviness. The waviness of one of the states of the hydrogen electron pictured above might then morph smoothly to another state without the quantum jumping Schrödinger detested.

That reasonable-seeming interpretation of waviness is wrong. Here's why: Though an object's waviness may be spread over a wide region, when one looks at a particular spot, one immediately finds either a *whole* object there, or *no* object in that spot.

For example, an alpha particle emitted from a nucleus might have waviness extending over kilometers. But as soon as a Geiger counter clicks, one can find a whole alpha right there inside the counter. Or consider the waviness of a single electron headed to the scintillation screen in the interference experiment that confirmed de Broglie's wave idea. Its waviness would be in several clumps, separated by inches. But an instant later a flash is seen at a *single spot* where the electron hits a screen. The whole electron could then be found there. The electron's previously extended waviness is suddenly concentrated at that single spot. If the electron were detected while in transit to the screen, it would be found concentrated at some single spot in one of the several clumps of its waviness.

If an actual physical object were smeared over the extent of its waviness, as Schrödinger initially thought, to fit the observed facts its remote parts would have to instantaneously coalesce to the place where the whole object was found. Physical matter would have to move at speeds greater than that of light. That's impossible.

In trying to rid physics of the "damn quantum jumps," Schrödinger failed. We will later find him objecting to something more outrageous than orbit-jumping electrons.

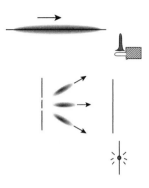

Figure 6.6 Top: Waviness of an alpha particle before and after detection by a Geiger counter. Bottom: Waviness of a single electron before and after detection on a screen

The Accepted Interpretation of Waviness

From the position and motion of objects at some time, Newton's laws of motion give their position and motion for all future times and all past times. From the wavefunction at some time, Schrödinger's equation gives the wavefunction for all future time, and for all past time. In that sense, quantum *theory* is as deterministic as classical physics. Quantum *mechanics*, the theory plus the experimental observations, has an intrinsic randomness. The randomness arises with "observation," something unexplained within the theory.

What we describe in the next few pages can be confusing. It's confusing because it's hard to *believe*. The accepted interpretation of waviness challenges any commonsense view of physical reality. It presents us with the quantum enigma.

The waviness in a region is the probability of *finding* the object in a particular place. We must be careful: The waviness is *not* the probability of the object *being* in a particular place. There's a crucial difference! The object was not there before you found it there. You could have chosen an interference experiment demonstrating it was spread out over a wide region. You know you could have done an interference experiment because that's what you actually did with other objects prepared in exactly the same way.

You could have made that choice in this case. Somehow, your looking *caused* it to be in a particular place. In our standard view of quantum mechanics, the Copenhagen interpretation (treated in chapter 10), "observations" not only disturb what is to be measured, observations actually *produce* the measured result. We'll later talk about what might be considered an "observation."

Waviness is probability, but we must contrast waviness, or quantum probability, with classical probability, something similar but intrinsically different. We start with an example of classical probability.

At a carnival, a fast-talking fellow with even faster hands operates a shell game. He places a pea under one of two inverted shells. After his rapid

shuffling, your eyes lose track of which shell holds the pea. There is equal probability for the pea to be in either of two places. We associate a probability of one-half with each shell, meaning that about half of the times we look we would find the pea under, say, the right-hand shell. The sum of the probabilities for the two shells is 1. (1/2 + 1/2 = 1). The sum of the two probabilities, a probability of 1, corresponds to the certainty that the pea is surely under one of the two shells.

Figure 6.7

After a bit of glib talk, as he takes some bets, the operator lifts the shell on the right. Suppose you see the pea. Instantaneously, it becomes a certainty that the pea was under the right-hand shell and that it is *not* under the left-hand shell. The probability "collapses" to zero for the left-hand shell and to 1, a certainty, for the right-hand shell. Even if that left-hand shell had been moved across town before the shell on the right was lifted, the collapse of probability would still be instantaneous. Great distance does not affect how fast probability can change.

Games of chance make it almost obvious what quantum waviness should represent. (Obvious at least to those of us who have previously been taught the answer.) It was, in fact, only a few months after Schrödinger announced his equation that Max Born put forward the now-accepted idea that the waviness in a region was *probability*, the probability for the *whole* object to be found in that region.

Figure 6.8

This Born postulate connects what we actually observe, a whole object in a particular place, with the mathematical expression of waviness that the quantum theory gives. Like probability in the shell game, when we find out where the object is, its waviness instantaneously "collapses" to 1 in the region we found it and to zero everyplace else.

There is, however, a crucial difference between the classical probability illustrated by the shell game and quantum probability represented by waviness. Classical probability is a statement of one's knowledge. In the shell game, your not knowing at all which shell covered the pea means that for *you* the probability of it being under each shell was 1/2. The shell-game operator likely had better knowledge. For *him* the probability was different.

Classical probability represents someone's knowledge of a situation. It is not the whole story. Something physical is presumed to exist in *addition* to that knowledge, something it was the probability *of*. There was a real pea under one of the shells. If someone peeked and saw the pea under the left-hand shell, the probability would collapse to a certainty, to 1, *for her*. But it could still be 1/2 for each shell for her friend. Classical probability is subjective.

Quantum probability, waviness, on the other hand, is mysteriously objective; it's the same for everyone. The wavefunction is the *whole* story: The standard quantum description has no atom in addition to the wave-function of the atom. As a leading quantum physics text would have it, the term "the wavefunction of the atom" is a *synonym* for "the atom."

If someone looked in a particular spot and happened to see the atom there, that look "collapsed" the spread-out waviness of that atom to be wholly at that particular spot. The atom would then be at that spot for everyone. (If he looked and found the atom *not* there, it would be not there for everyone.) If that someone observed the atom at a particular spot, a second observer looking at a different spot would surely not find the atom at that different spot. Nevertheless, the waviness of that atom existed at that different spot immediately before the first observer collapsed it. Quantum theory insists this is so because an interference experiment *could have* established the waviness of that atom to have existed there. (This is admittedly confusing. The situation will get clearer when we describe the experiments leading to these conclusions. But an enigma will remain.)

Observing an atom being at a particular place *created* its being there? Yes. But we must be careful here. We're touching on something controversial: "observation." The standard view (or the Copenhagen interpretation, sometimes called physics' "orthodox" view) considers that an observation takes place whenever a small, microscopic, object affects a large, macroscopic, object. Should an atom cause a flash someplace on a scintillation screen, for example, that macroscopic screen, in the Copenhagen interpretation, "collapses" that atom's widely spread-out wavefunction to be concentrated at that spot on the screen.

However, just before the atom hit the screen, it was a widely spread-out wave. Hitting the screen, it somehow became a particle concentrated at a particular spot. We could look and find it there. We can therefore say that the screen has "observed" the atom. This is a pretty good way to go, at least for all *practical* purposes. We will, however, be interested in what's going on *beyond* mere practical purposes.

We've been talking of an atom because quantum theory was developed to deal with microscopic objects. But quantum theory is basic to all of physics, all of science, and is applied to entities as large as the universe, and as intimate as the mind, though doing so is controversial.

Intrinsically Probabilistic

A theory in physics predicts what you will see in an experiment, where an "experiment" is any well-specified situation. For a tossed ball, or a planet, classical physics tells the actual position of the ball or planet at any time, *even when it is not being observed*. There might be uncertainty in such predictions, which may specify a range of possible positions. Though the predictions may be probabilistic, the object is assumed to actually exist at some particular place. In classical physics, probability is the subjective uncertainty of our knowledge.

Quantum mechanics, on the other hand, is *intrinsically* probabilistic. Probability is all there is. Quantum physics does not tell the probability of where an object *is*, but rather the probability that, if you look, you will *observe* the object at a particular place. The object has no "actual position" before that position is observed. In quantum mechanics the position of an object is not independent of its observation at that position. The observed cannot be separated from the observer.

Let's look at two attitudes about quantum probability.

There is the "It's all OK!" attitude: Waviness is the probability of what you will *observe*. Yes, it depends on how you look. You can directly look at an object and demonstrate it to be a compact thing in a particular place. Or you can do an interference experiment and demonstrate it had been a widely spread-out thing. In either case, quantum mechanics predicts the correct outcome for the experiment you actually do. Since correct predictions are all one ever needs, for all practical purposes there is no problem. We defend this useful pragmatic attitude, the Copenhagen interpretation, in chapter 10.

On the other hand, there is the "I'm baffled!" attitude: The theory only gives waviness. This is an *assumption* that goes *beyond* the Schrödinger equation. It's the Born postulate that tells us that observation collapses the spread-out waviness to the particular place we happen to find the object.

Does Nature's fundamental law, the Schrödinger equation, provide only *probability*? Einstein felt that there must be an underlying deterministic explanation for the particular position at which the object was found: "God does not play dice." (Bohr suggested Einstein not tell God how to run the universe.)

Randomness was *not* Einstein's serious problem with quantum mechanics, despite this much-quoted theological comment. What disturbed Einstein, and Schrödinger, and more experts today, is quantum mechanics' apparent denial of physical reality. Or, maybe the same thing, that the observer's *choice* of how to observe affects the *prior* physical situation. According to quantum theory, there was *not* an actual atom in a particular place before we looked, "collapsed the wavefunction," and *found* an atom there. But there *are* actual atoms, and actual things made of atoms. Aren't there?

In the early 1920s, before the Schrödinger equation, the fact that one could display light and matter *either* as a spread-out wave *or* as a collection of compact particles was a troubling puzzle. However, it was hoped that some yet-to-be-found fundamental theory might give a reasonable explanation. By the late 1920s, with the Schrödinger equation, the fundamental theory seemed in hand. But the puzzle was even more troubling.

The Two-Slit Experiment
The Observer Problem

[The two-slit experiment] contains the only mystery. We cannot make the mystery go away by "explaining" how it works . . . In telling you how it works we will have told you about the basic peculiarities of all quantum mechanics.

—Richard Feynman

In this chapter, we go for rigor in presenting the quantum enigma. In the rest of the book, we more loosely ponder what it all might mean.

The two-slit experiment, the archetypal demonstration of quantum phenomena, displays physics encounter with consciousness. Quoting Feynman above, "We cannot make the mystery go away...." But we'll tell how it works.

The two-slit experiment is, in part, an interference experiment. We described interference for light waves in chapter 4. Interference has been demonstrated with photons, electrons, atoms, and large molecules, and is being attempted with yet bigger things. Demonstrating interference with photons is an easy classroom demonstration. The slits can be two lines scratched on an opaque film. Shining a laser pointer through the slits, you can display a clear interference pattern. Electron interference is not so easy, but you can buy a dramatic classroom demonstration apparatus for a few thousand dollars. Demonstrating interference with atoms or molecules is trickier and far more expensive. But it's basically the same idea. Since electrons or atoms would collide with air molecules, interference with objects other than photons must be displayed in a container from which the air is removed, but we'll not worry about such technical "details."

Since the quantum mystery is the same in every case, and since talk presents no budget problems, we will talk in terms of atoms. Today we can see individual atoms, even pick them up and put them down one at a time. We first briefly describe the standard version of the two-slit experiment. We then follow it with a completely equivalent version that contrasts nicely with the shell game of chapter 6.

In describing the interference of light waves in chapter 4, we noted that to get a well-defined interference pattern, the light should be of a single color. That means light of a narrow range of frequencies and wavelengths. The same applies for atoms. The atoms should all have essentially the same de Broglie wavelength, which just means they should all come with the same speed.

Our slits are the two openings as shown in figure 7.1. You send in atoms from the left. Coming through the slits, they land on a screen to the right,

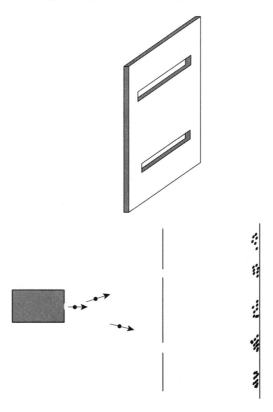

Figure 7.1 Top: The two-slit diaphragm. Bottom: Edge-on view of atom source, two-slit diaphragm, and detection screen with atoms in interference pattern.

which we show in figure 7.2. (We don't care about any atoms that fail to come through the slits.)

You record where on the screen the atoms land. They land only in certain regions. The distribution of atoms yields the pattern shown in figure 7.2. (It's the same as our figure 5.7 for light waves.)

The pattern, an interference pattern, comes about, as with any waves, because each atom's wavefunction comes through both slits. At some places on the screen, crests from the top slit arrive together with crests from the bottom slit. Waves from the two slits then add to produce regions of large waviness. Elsewhere, crests from one slit arrive with troughs from the other to cancel, producing regions of zero waviness. The waviness someplace is the probability of finding an atom there. You thus find regions where many atoms have hit and regions where few atoms have hit. In the "orthodox" Copenhagen interpretation of what's going on, the wavefunction of each atom collapsed at the particular point where it hit, where it was "observed" by the macroscopic screen.

Figure 7.2 Interference pattern formed by atoms coming through two narrow slits

Since each atom's wavefunction followed a rule that depended on the spacing of the slits, *something* of each atom *must* have come through both slits. Quantum theory has no atom in addition to the wavefunction of the atom. Accordingly, each atom itself must have been a *spread-out* thing coming through both of the well-separated slits.

However, you *could have* done this experiment with only one slit open. Each atom's wavefunction could then have come through only a single one of the narrow slits. You still find atoms landing on the screen. There could, of course, be no interference because each atom's wavefunction came through only a single slit. But since each atom's wavefunction came through a *single* narrow slit, each atom is established to be a *compact* thing, a particle. The atoms fall in the uniform distribution shown in figure 7.3.

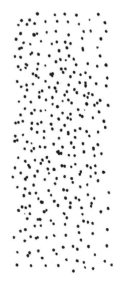

Figure 7.3 Distribution of atoms coming through a single narrow slit

You thus could choose to demonstrate, with both slits open, that atoms are spread-out things. Or, with only a single slit open, you could choose to demonstrate the opposite, that atoms are compact particles. This is, of course, the wave-particle paradox discussed for de Broglie waves in chapter 5. We just told the story for atoms in terms of today's quantum theory.

Our Box-Pairs Version of the Two-Slit Experiment

Here's a completely equivalent version of the two-slit experiment in which you can choose to show that an object, an atom, for example, was *wholly* in a single box. But you *could have* chosen to show that that same atom was *not* wholly in a single box. Telling the story with atoms captured in boxes, you can decide *at your leisure* which contradictory situation you wish to demonstrate. This way of telling the story more dramatically displays the quantum challenge to our commonsense intuition that an observer-independent physical reality exists "out there." We'll refer to our box pairs again—*and again*—in future chapters. So we tell it carefully here.

Aristotle taught that to discover Nature's laws one should start with the simplest examples, and from them move on to greater generalizations. Galileo accepted that injunction, but warned that we must rely on only what is *experimentally* demonstrable, even if the results violate our deepest intuitions. Considering the idealized behavior of isolated objects, the moon, the planets, and apples, Newton formulated his universal equation of motion. The two-slit experiment, the simplest display of quantum phenomena, follows this path. We carefully treat our box-pairs version with atoms. We later generalize it to cats, to consciousness, and to the cosmos.

In the shell game of chapter 6, the pea had equal probability for being under each shell. Probability was not the complete description of the physical situation. There was also an actual pea definitely under one shell or the other. Observation did *not* change that physical situation. We will put equal parts of the waviness of a single atom in each of two boxes, so that the atom has equal probability for being in either box. But, unlike the shell game, there is no "actual atom" in a particular box. The wavefunction divided into both boxes is the *complete* description of the physical situation. And here, unlike the shell game, observation *does* change the physical situation.

To display the quantum enigma, it is *not* necessary to tell how our box pairs are prepared. However, since we've already spoken of wavefunctions, we will describe the preparation. After that, however, we will display the quantum enigma telling *only* what you would actually *see*. We will describe the box-pairs experiment, without mentioning quantum theory, or wavefunctions, without even mentioning waves.

Here's how the atoms were put into the box pairs. Any wave can be reflected. A semitransparent mirror reflects part of a wave and allows the rest to go through. A glass windowpane, for example, allows some light through and reflects some. At the glass, the wavefunction of each individual photon splits. Part of the photon wavefunction is reflected and part is transmitted. We can also have a semitransparent mirror for atoms. It splits an atom's wavefunction into two wave packets. One packet goes through, and another is reflected.

The arrangement of mirrors and boxes in figure 7.4 allows for the trapping of the two parts of an atom's wavefunction in a pair of boxes. We send in a single atom at a known speed and close the doors of the boxes when the wavefunction packets are inside the boxes. After that, each part of the wavefunction bounces back and forth in its box. In figure 7.4 we show the wavefunction and waviness at three successive times.

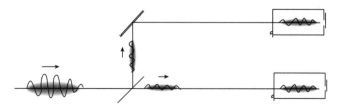

Figure 7.4 Mirror and box-pair setup allowing the trapping of a wavefunction in a pair of boxes. An atom's wavefunction is shown at three different times.

We know there is one, and only one, atom in each box pair because we *observed* an atom and sent one into each box pair. These days, with the proper tools, we can see and deal with individual atoms and molecules. With a scanning tunneling microscope, for example, we can pick up and put down, single atoms.

Holding an atom in a box without disturbing its wavefunction would be tricky, but it's certainly doable. Dividing the wavefunction of an atom into well-separated regions is accomplished in every actual interference experiment with atoms. Capturing the atoms in physical boxes is actually not needed for our demonstration. A defined region of space would be enough. We like to think of each region defined by a box because it's more like the shell game. We can then consider the atom sitting there waiting for us to choose what to do with it, rather than have the atom zip through a two-slit diaphragm on its way to the detection screen.

From here on, the description of our box-pairs experiment will not mention wavefunctions, or waves at all. We just tell what you would actually see. We describe *quantum-theory-neutral* observations. By doing this we emphasize that the quantum enigma arises *directly* from *experimental* observations. The existence of the quantum enigma does *not* depend on the quantum *theory*!

The "Interference Experiment"

You are presented with large number of box pairs. (They were prepared as we described above, but for the demonstration of the enigma, you need not know anything of the preparation.) Position a box pair in front of a screen on which an impacting atom would stick. Open a narrow slit in each box, at about the same time. An atom hits the screen. Repeat this with many identically positioned box pairs. You find that atoms cluster in some regions of the screen, but avoid other regions. The pattern is the same as that shown previously in figure 7.2 for a pair of slits. Each atom followed a rule allowing it to land in certain regions and forbidding it from landing in other regions.

Now repeat this procedure with a new set of box pairs. This time have a different spacing between the boxes of each pair. You find the regions where the atoms clustered are *spaced* differently. The larger the spacing between the boxes of a pair, the smaller is the spacing between the places

Figure 7.5 Interference patterns formed by atoms coming through two narrow slits with different slit separations

where atoms land. We illustrate this with figure 7.5. Each and every atom followed a rule that *depends on the spacing* of its box pair. Each atom therefore had to "know" its box-pair spacing.

Clearly, the experiment we just described is an interference experiment, like the two-slit experiment, and we'll now call it an "interference experiment." But we *did not* use any property of waves. Something of each atom had to come from each box because where atoms landed depended on the box-pair spacing. This interference experiment establishes that each atom had been a spread-out thing, in both boxes of its pair. (Nothing done *outside* the boxes while the atom is still inside has any effect at all.)

What explains a *whole* atom appearing on the screen while some of it had to come from *each* box of its pair? Might it not make sense to say that *part* of each atom *was* in each box? In that case, part of the atom emerged from each box of its pair, and then congealed to the spot on the screen where you found it. That reasonable-sounding idea doesn't work. Here's why:

The "Which Box?" Experiment

Instead of doing the experiment by opening the slits in the boxes of a pair at the *same* time, choose a different experiment: Open a slit in one box, and then *later*, open a slit in the other box. Opening one box, you sometimes find a *whole* atom impacts the screen. If so, when you open the other box of that pair, *nothing* comes out. If you open a box and nothing appears on the screen, then, for sure, an atom will appear on the screen when you open the second box. Repeatedly opening boxes of a set of box pairs one box at a time, you determine which box the *whole* atom was in. You demonstrate that there had been a *whole* atom in one box, and that the other box of that pair contained *nothing*. With atoms wholly in a single box, box spacing would not be relevant. Indeed, you find a uniform distribution of atoms hitting the screen, as previously seen in figure 7.3 for the case of a single slit being open.

There's a more direct way to establish that each atom was *wholly* in a single box. Simply *look* in a box to see which box held the atom. It doesn't matter how you look. You can, for example, shine an appropriate light beam into the box and see a glint from the atom. About half the time you will find a whole atom in the looked-in box; about half the time you find the box empty. If there is no atom in the box you first look in, it will always be in the other. If you find an atom in one box, the other box of its pair will be *totally empty*. No observation, of any kind, would find anything at all in that empty box. This "which-box?" experiment, or "look-in-the-box" experiment, establishes that each atom was concentrated in a *single* box of its pair, that it was *not* spread out over both boxes.

But *before* you looked, you could have done an interference experiment establishing that something of each atom had been in *both* boxes. You therefore could choose to prove either that each atom had been *wholly* in a single box, or you could choose to prove that each atom had *not* been wholly in a single box. You can choose to prove either of two *contradictory* situations.

The ability to prove *either* of two contradictory results is puzzling. Wanting to explore further, some have asked: "What if you do *both* experiments with the same atoms? What if you open the box pairs at the same time, in order to get an interference pattern, but *also* look to see which box each atom came out of." Such looking is essentially a which-box experiment. Absolutely *anything* you do that allows you to know which box the atom was in defeats the atom's ability to obey the rule giving an interference pattern.

Seeking a loophole, a logician might note that the interference experiment relies on *circumstantial* evidence. It uses one fact, the interference pattern, to establish another fact, that each atom came out of both boxes.

This is true of *any* interference experiment. Finding no other reasonable explanation, physics universally accepts interference as establishing spread-out waviness. As in our legal system, circumstantial evidence can establish a conclusion beyond a reasonable doubt.

A theory leading to a logical contradiction is necessarily an incorrect theory. Does the ability to demonstrate either of two contradictory things about atoms (and other objects) invalidate quantum theory? No. You did not demonstrate the contradiction with *exactly* the same things. You did the two experiments with different atoms.

The Quantum Enigma

Here's a logically *conceivable* explanation of your ability to prove either of two contradictory things: Those box pairs for which you chose an interference experiment actually *contained* objects spread out over both boxes, *not* wholly in a single box. And those box pairs for which you chose a which-box experiment actually *contained* compact objects wholly in a single box. How could you establish otherwise?

You reject this explanation. You reject it because you *know* that, given a set of box pairs, you could have made *either* choice. You freely *chose* which experiment to do. You have free will. At least your choices were not predetermined by a physical situation external to your body, by what was supposedly "actually" in the box pairs.

Did your free choice determine the external physical situation? Or did the external physical situation predetermine your choice? Either way, it doesn't make sense. It's the unresolved quantum enigma.

Important point: We experience an enigma because we believe that we *could have* done other than what we actually did. A denial of this freedom of choice requires our behavior to be programmed to correlate with the world external to our bodies. The quantum enigma arises from our *conscious perception* of free will. This mystery connecting consciousness with the physical world displays physics' encounter with consciousness.

History Creation

To a certain extent at least, our present actions obviously determine the future. But obviously, our *present* actions cannot determine the *past*. The past is the "unchangeable truth of history." Or is it?

Finding an atom in a single box means the whole atom came to that box on a particular *single* path after its earlier encounter with the semi-transparent mirror. Choosing an interference experiment would establish a *different* history: that aspects of the atom came on *two* paths to *both* boxes after its earlier encounter with the semi-transparent mirror.

The creation of past history is even more counterintuitive than the creation of a present situation. Nevertheless, that's what the box-pairs experiment, or any version of the two-slit experiment, implies. Quantum theory has *any* observation creating its relevant history. (We'll see this dramatically displayed with Schrödinger's cat story.)

In 1984, quantum cosmologist John Wheeler suggested quantum theory's history creation be put to a direct test. He would have the choice of which experiment to do delayed until *after* the object made its "decision" at the semi-transparent mirror: whether to come on a single path or whether to come on both. Atoms in box pairs would be too difficult for a practical experiment. It was done with photons and a mirror arrangement much like our figure 7.4. Getting the same results as in the usual quantum experiment would imply that the relevant history was indeed created by the later choice of which experiment to do.

For a human to make a conscious choice of which experiment to do takes perhaps a second. But in a second, a photon travels 186,000 miles. We can't build an apparatus this big, or have a photon bounce back and forth in a box for as long as a second. In the actual test, the "choice" of experiment was therefore made by a fast electronic switch driven by a random number generator. The most rigorous version of the experiment was not done until 2007, when reliable single-photon pulses could be generated, and fast enough electronics were available. The result was (of course?) that the quantum theory predictions were confirmed. Observation created the relevant history. In Wheeler's words: "We have a strange inversion of the normal order of time . . . an unavoidable effect on what we have a right to say about the already past history of that photon."

The Enigma Is Displayed *Experimentally*

In our box-pairs experiment, we described only what you would actually see. We never referred to quantum *theory*. Contrast this ***quantum*** enigma with the enigma of Newtonian determinism. Taken to its logically extreme conclusion (as is sometimes done), Newtonian determinism denies the possibility of free will. However, this Newtonian enigma arises only from the deterministic Newtonian *theory*. Classical physics predicts no **experimental** consequences challenging the belief that our free choice can arise wholly within our body.

The quantum enigma, on the other hand, arises directly from experiment. It's harder to ignore an enigma arising directly from experimental observation than one arising only from theory.

If the quantum enigma is independent of quantum theory, why do we call it a "*quantum* enigma"? Because theory-neutral experiments, like the two-slit experiment, form the *basis* of quantum theory. Quantum theory provides a mathematical description that correctly predicts the results of the experiments, of the observations that we *choose* to make.

The Quantum *Theory* Description

Now that we have established the *experimental* basis of the enigma, we offer quantum theory's explanation. Since we can choose to observe an atom to be in either of two contradictory situations, how does quantum theory describe the state of the atom *before* we observe it? The theory describes the world in mathematical terms. In those terms, when an atom can be *observed* in either of two contradictory situations, or "states," the wavefunction of the total physical situation is written as the *sum* of the wavefunctions of those two states separately. Expressing this mathematics in words, the wavefunction of one of the states is "the-atom-is-wholly-in-the-top-box." The wavefunction of the other state is "the-atom-is-wholly-in-the-bottom-box." The wavefunction of the unobserved atom is "the-atom-is-wholly-in-the-top-box" plus "the-atom-is-wholly-in-the-bottom-box." The atom is said to be in a "superposition" of these two states. It is simultaneously in both states. On looking in a box, this sum, or superposition, collapses randomly to one or the other term of the superposition. But before we look, the atom is simultaneously in both boxes. The atom is in two places at once.

Observation collapses the waviness, the probability, to a specific actuality. But what constitutes an "observation"? Observation is ultimately not explained within quantum theory. What constitutes observation is controversial. The pragmatic Copenhagen interpretation of quantum mechanics, physics' "orthodox" position (more fully discussed in chapter 10) defines any recording of a microscopic event by a macroscopic measuring instrument as an observation. Or, more strictly, *any* interaction of a microscopic system with a macroscopic system constitutes an observation if it would make a demonstration of interference essentially impossible. Not all physicists accept this for-all-practical-purposes interpretation of observation. We leave the issue, for now. We can, however, tell what all physicists would agree does *not* constitute an observation.

When one microscopic object encounters a second microscopic object, does the first object "observe" the second? No. As an example, consider an atom in our box pair, simultaneously in two boxes. Say a photon is sent through the (transparent) top box. Should the atom actually *be* in that box, the photon would be deflected. Were the atom instead in the bottom box, the photon would come straight through that top box. Did the photon "observe" whether or not the atom was in the top box? No. The photon entered a superposition state with the atom. We say it "entangled" with the atom. A rather complicated interference experiment could actually establish that the entangled atom–photon system was in a state in which the photon was *both* deflected by the atom and *not* deflected by the atom.

When that simultaneously deflected and not-deflected photon later encountered other objects, anything macroscopic, say a Geiger counter, no demonstration of interference would be possible, for all practical purposes. We can then consider an observation to have been made, the wavefunction to have collapsed. Seeing whether or not the Geiger counter fired, we could surely say whether or not the photon scattered from the atom, and thus whether or not the atom was in the top box.

We've emphasized a quantum enigma arising from quantum-theory-neutral experimental observations. One can see a different enigma arising from the quantum theory. Theory has the atom in our box pairs in a superposition state with waviness equally in both boxes. But on looking, we find the atom wholly in a single box. How does Nature decide on a *particular* result, a *particular* box, when quantum theory, our most basic description of Nature, gives only probability?

It's unexplained. There's an intrinsic randomness associated with observations. Choosing to observe the atom wholly in a single box, we cannot choose in *which* box it will appear. Or choosing an interference experiment, we cannot choose in *which* of the allowed regions the atom will appear. We can choose the game, but not the particular outcome. Wavefunctions collapse with a randomness. (Pseudo-scientific invocations of quantum mechanics can ignore this randomness to imply that your choice of thoughts alone can bring about a particular desired result.)

We have displayed an object's *position* being created by observation. Observation-creation applies to every other property as well. For example, many atoms are tiny magnets with a north pole and a south pole. The atom can be put into a superposition state with its north pole simultaneously pointing both up and down at the same time. But an observation of its orientation always yields either up *or* down.

Though we've just been talking of an atom, quantum theory presumably applies to everything. We later come to Schrödinger's story logically extending this reasoning to the *impossible-in-practice,* but logically-consistent-with-quantum-theory, situation of a cat in two contradictory states, alive and dead at the same time. Something being in two mutually exclusive states at the same time is confusing. Some confusion will be straightened out in later chapters. But not all of it! We have confronted the still unresolved, and definitely controversial, quantum enigma. However, the *experimental* results we have described are completely undisputed.

In our next chapter, we consider these same ideas in a lighter vein.

8

Our Skeleton in the Closet

The interpretation [of quantum mechanics] has remained a source of conflict from its inception. . . . For many thoughtful physicists, it has remained a kind of "skeleton in the closet."

—J. M. Jauch

In his book *Dreams of a Final Theory*, Nobel Laureate Steven Weinberg writes: "The one part of today's physics that seems to me likely to survive unchanged in a final theory is quantum mechanics." We share Weinberg's intuition about the ultimate correctness of quantum mechanics.

John Bell, a major figure in our later chapters who would likely have the Nobel Prize if it could be awarded posthumously, felt that "the quantum mechanical description will be superseded. . . . It carries in itself the seeds of its own destruction." Bell does not really disagree with Weinberg. His concern with quantum theory is not that an error will be found in any of its predictions, but that it is not the whole story. For him, quantum mechanics reveals the incompleteness of our worldview. He feels it is likely "that the new way of seeing things will involve an imaginative leap that will astonish us." (Incidentally, Bell tells that it was a lecture by Jauch—whom we quote just above—that inspired his investigations into the foundations of quantum mechanics.)

Along with Bell, we suspect that something beyond ordinary physics awaits discovery. Not all physicists would agree. Many, if not most, would minimize the enigma, as something we should just get used to. It's our "skeleton in the closet,"

However, the existence of an enigma is not a physics question. It's *metaphysics* in the original sense of that word. (*Metaphysics* is the name of Aristotle's work that followed his scientific text *Physics*. It treats more general philosophical issues.) When it comes to metaphysics, non-physicists with a general understanding of the experimental *facts*—facts about which there is no dispute—can have an opinion with a validity matching that of physicists.

We illustrate this point with a story in which an orthodox-minded physicist demonstrates the basic experimental facts of quantum mechanics (described in our previous chapter) to a Group of Rational and Open-minded PEople (the GROPE) who have never been exposed to the quantum theory that explains those facts. What our physicist demonstrates to the GROPE is analogous to the experience of the visitor to Neg Ahne Poc. Though what was displayed in Neg Ahne Poc is *not* actually possible, that visitor's bafflement is the same bafflement the GROPE experiences from a demonstration that *is* actually possible. You may share that bafflement. We do. It's the quantum enigma.

After her demonstration, our physicist offers the standard quantum theory explanation for what was seen. It's the explanation that generally satisfies students in quantum physics classes. Their concern with the calculations that will be on their exams overrides their interest in the *meaning* of what they calculate. The GROPE, on the other hand, is concerned with what it all might *mean*. In discussing the enigma, we hope you can identify with the GROPE. We can.

The "apparatus" our physicist uses is a caricature of an actual laboratory setup. But the quantum phenomena she demonstrates are well established for small objects. These phenomena are today being displayed with ever-larger objects. Midsized proteins, and even viruses, are the objects in present experiments. Quantum theory sets no limit. The size of objects shown to exhibit such quantum phenomena seems constrained only by technology, and budget.

We could be completely general in our story and talk of the experiments being done with "objects." That sounds vague. There's no reason we can't think of our objects as little green marbles. The experiment could actually be done with "little green marbles," as long as they were *very* little marbles, say, the size of large molecules. So for our story, we talk in terms of "marbles."

Our physicist warmly welcomes the GROPE, telling them, "I've been asked to demonstrate to you the strange nature of 'observation' and to tell quantum theory's explanation of what you will see. Sometimes we physicists hesitate to call attention to this strangeness because it can make physics seem mystical. But I'm assured that you're a group of rational, open-minded people for whom that's not a problem. I believe I can show you something truly remarkable."

The first experiment our physicist does should remind you of the visitor in Neg Ahne Poc asking: "In which hut is the couple?" The appropriate answer he always got demonstrated the couple to be wholly in one hut or the other.

Our physicist points to a set of boxes, each box paired with another. She explains that her apparatus injects a single marble into each *pair* of boxes. "Just how my apparatus works," she says, "won't matter at all for my demonstration." The GROPE accepts this. They watch as she mounts a box pair on the right end of her apparatus, drops a tiny marble into a hopper on the left, and then removes the box pair. She repeats the procedure, accumulating a few dozen box pairs.

Unlike the GROPE, you have been exposed to quantum theory. We therefore note that our physicist's apparatus involves a set of mirrors appropriate for dividing the waviness of each "marble" equally into both boxes of each pair.

Figure 8.1

🚶 "My first experiment," our physicist explains, "will determine which box of each pair contains the marble." Pointing at a box pair, she nods to one eager-looking member of the GROPE and asks: "Would you please open each box and see which box holds the marble?"

Opening the first box, the young man announces: "Here it is."

🚶 "Make sure the other box is completely empty," requests our physicist.

Looking carefully, he says with assurance, "It's completely empty—there's nothing in it."

Once he was through examining the boxes, our physicist asks an attentive young woman to repeat the procedure of finding which box of another pair held the marble. Opening the first box, she remarks, "It's empty; the marble must be in the other box." Indeed, she finds it there.

Our physicist repeats this procedure several more times. The marble appears randomly, sometimes in the first box opened, sometimes in the second. She soon notices members of the GROPE not paying much attention and mumbling to each other. She overhears one fellow say to the woman next to him: "What's her point? Hardly the remarkable demonstration we were promised."

🚶 Though the remark was not directed at her, our physicist responds: "I'm sorry, I just want to convince you that when we look to find out which box of a pair holds the marble, we demonstrate that there is

a whole marble in one box and that the other is completely empty. Please bear with me, because I'd now like to show you that it doesn't matter just *how* we find out in which box our marble is. Here's another way to find out."

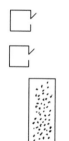

She sets a box pair in front of a sticky screen and opens one box. One can't see the fast-moving marble, but there is a "plink," and a marble sticks to the screen. "Ah, the marble was in the first box," she says. "Therefore, no marble will hit the screen when I open the second box."

Figure 8.2
Opening boxes
sequentially,
with results on
the screen

"Obviously," is a mumbled comment from someone near the back of the GROPE.

Though holding the attention of the GROPE again becomes difficult, our physicist repeats the demonstration with more box pairs. If a marble hits the screen when she opens the first box, none appears when she opens the second box. If no marble appears on the screen on the first opening, there is always a marble on the second. The screen gradually becomes spotted with marbles, distributed more or less uniformly over the screen.

"Can you see," she asks, "that this is another demonstration that there is a marble in one of the boxes of a pair and that the other is empty?"

"Sure, but where's the *remarkable* demonstration you promised?" grumbles one fellow. "Of course *how* you look doesn't matter. Your apparatus put a marble in one box of each pair. So what?" Several nod agreement. And from an outspoken woman: "He's right!"

"Actually," our physicist says hesitantly, "the remarkable thing—what I hope to demonstrate—is that what he just said is *not* quite right. Let me try another experiment first."

The next experiment our physicist does should remind you of the visitor to Neg Ahne Poc asking: "In which hut is the man and in which hut is the woman?" The appropriate answer he always got demonstrated the couple to be distributed over both huts.

The GROPE politely settles down to watch the new experiment.

Our physicist positions a new set of box pairs in front of the sticky screen and quickly opens *both* boxes of the pair. "The difference in this next experiment," she points out, "is that I'm opening both boxes at about the same time." A plink indicates the impact of a marble on the screen. Discarding that box pair, our physicist carefully positions another in the same place and again opens both boxes together. Another plink is heard as a marble hits the screen.

Marbles accumulate on the screen as she opens more box pairs simultaneously. A fellow in a red shirt asks idly: "Doesn't this experiment demonstrate even less than your first one? Since you're now opening both boxes at the same time, for this set, we can't even tell which box the marble came out of."

But before his remark is seriously considered, a previously silent woman up front says: "Where the marbles land seems to form a pattern."

Figure 8.3
Results on
screen of
opening boxes
simultaneously

Now they all watch carefully. As more marbles plink onto the screen, the pattern emerges distinctly. Marbles land only in certain places. In other places on the screen there are no marbles. Each marble follows a rule allowing it to land only in certain places and forbidding it to land in others.

The woman who first noticed the pattern seems puzzled and now asks: "In your first experiment, when the boxes of each pair were opened separately, the marbles were uniformly distributed over the screen. How can opening the *empty* box along with the one holding the marble affect where the marbles land?"

 Our physicist, delighted with that question, responds eagerly: "You're right! Opening a box that was truly empty couldn't have any affect. I told you that there was a marble in each *pair* of boxes. But it's not right to say that one box held the marble and the other was empty. Each and every marble was simultaneously in both boxes of its box pair."

Responding to the dubious looks on the faces of several members of the GROPE, our physicist persists: "I know this is hard to believe, but there's a quite convincing way to show that. It's just a bit time-consuming."

The GROPE chats and relaxes as our physicist and her graduate student assistant quickly prepare three sets of box pairs, each set containing a dozen or more box pairs. Now, regaining the GROPE's attention, she repeats her simultaneous openings of both boxes of each pair. But this time with each of the three sets of box pairs she uses a different spacing for the boxes of the pair.

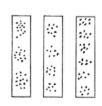

Figure 8.4 Results
of opening boxes
simultaneously
with different box
spacings

"Notice that the farther apart the boxes of a pair are, the closer spaced is the pattern. The rule that *each and every* marble obeys, the rule that tells each marble the places where it is allowed to land, depends on the *spacing* of its box pair. Each marble therefore 'knows' that spacing. Each marble occupied both boxes of its pair."

"Wait a second, lady," a kid pipes up. "You're saying the marble was in *two places at the same time*, that it came out of *both* boxes. That's silly! . . . Ah, oh, I'm sorry, ma'am."

"No problem, young fellow," responds our physicist. "You're quite right. The marble was simultaneously in two places. It was in *both* boxes. The scientific way is to accept what Nature tells us regardless of our intuitions. A single marble coming out of both boxes may sound silly, but our experimental observations leave us little alternative."

This takes a bit of contemplation. But after a minute or so, that fellow in the red shirt again speaks up: "There *is* an alternative, an *obvious* one. In your first experiment, where you opened boxes one at a time, we saw one box of each pair to be completely empty. But, as you just said, for these other sets of box pairs, the marbles were split so that *something* of each marble went into both boxes of its pair. Clearly these sets of box pairs were *prepared* differently."

 Our physicist pauses with her hands on her hips to allow this idea to take hold before she comments: "That's a reasonable hypothesis. But actually all of the box pairs were prepared identically. Let me convince you that the preparation is not the issue with some box-pair sets we'll prepare."

This third experiment our physicist does should remind you of the visitor to Neg Ahne Poc asking either question. He could freely choose to demonstrate

Figure 8.5 Drawing by Charles Addams. © Tee and Charles Addams Foundation

Figure 8.6

either that the couple had been in a single hut or that the couple had been distributed over both huts. That baffled him.

After their coffee break, during which our physicist and her assistant prepared and stacked up several sets of box pairs, the GROPE reassembles. A puzzled man speaks up: "We've been talking about what you said, and at least some of us are confused. A few of us thought you claimed to demonstrate that one box of each pair was *empty*. But then you later claimed that *neither* box was empty. Those are two contradictory situations. They misunderstood you. Didn't they?"

"Well, they have it almost right. Which situation would you like to demonstrate with *this* group of box pairs?"

Somewhat taken aback, the questioner hesitates, but the woman next to him quickly volunteers: "OK, show us that for this set one box of each pair is empty."

Our physicist repeats her first experiment, opening the boxes of each of a dozen box pairs in turn. Each time she reveals a marble in one of the boxes and shows the other box to be empty. She comments: "And I assure you that no matter how the empty boxes are investigated, absolutely nothing would ever be found in them."

A cooperative fellow now points to another set of box pairs and asks: "Can you now show us that for this other set of box pairs *neither* box is empty?"

"I can do that," she says, and our physicist, opening both boxes of each pair simultaneously for a dozen box pairs, performs her second experiment, demonstrating that each marble must have occupied *both* boxes of its pair.

Several times our physicist demonstrates *either* of the two apparently contradictory situations, as chosen by a GROPE member.

A fellow up front brusquely calls out in the middle of one of the demonstrations: "What you're telling us—and I admit seem to demonstrate—makes no sense. It's logically inconsistent. . . . Oh, I apologize, I didn't mean to interrupt."

"No, no, it's okay," our physicist assures him. "You raise an important point."

He therefore continues: "You claim to demonstrate that both boxes of each pair contain at least something of the marble, but you supposedly also show that one box of each pair is totally empty. That's logically inconsistent."

"You'd be right," replies our physicist, "if we showed both those results for the *same* set of box pairs. But since we actually did those two demonstrations with two *different* sets of box pairs, there's no *logical* inconsistency."

A woman objects: "But for the box pairs for which you demonstrated one thing, we *could* have asked you to demonstrate the opposite."

"But you didn't," is our physicist's almost too casual reply. "Predictions for not-done experiments can't be tested. Therefore, *logically*, there's no need to account for them."

"Oh, no, you can't squeeze through that loophole," the original objector retorts. "We're conscious human beings, we have free will. We *could* have made the other choice."

Our physicist squirms a bit: "Consciousness and free will are really issues for philosophy. I admit these issues are raised by quantum mechanics, but most of us, most physicists, prefer to avoid such discussion."

An earlier questioner is unsatisfied: "OK," he demands, "but you agree that before we looked, one box of each pair in fact had a marble in it, or was empty. You physicists believe in a physically real world, don't you?"

He considered his question rhetorical. At least he expected a "Yes, of course" answer.

But our physicist hesitates, and again seems evasive: "What existed before we looked, what you call 'a physically real world,' is another issue most physicists prefer to leave to philosophers. For all practical purposes, all we need deal with is what we see when we actually *do* look."

"But you're saying something crazy about the world!" the questioner exclaims. "You're saying that what previously existed is created by the way we look at something." Most heads nod in agreement; others just seem baffled.

🧍 "Hey, I promised I'd show you something remarkable. I've done that, haven't I?" Responding to some nods, but more frowns, she continues: "We find the world stranger than we once imagined, perhaps stranger than we *can* imagine. But that's just the way it is."

"Wait!" a previously silent woman says firmly. "You can't get away with avoiding the issues your demonstrations raise. There's got to be an explanation. For example, instead of being in both boxes, maybe every marble has a kind of undetectable radar that tells it the separation of its box pair."

🧍 "We can never rule out 'undetectable' things," our physicist admits. "But a theory with no testable consequences beyond what it was invented to explain is unscientific. Just as useful as your theory of an 'undetectable radar' would be to assume that an invisible fairy rides on and guides each marble." Realizing she has embarrassed the proposer of the radar theory, our physicist apologizes: "I'm sorry; that was snide. Speculations like yours can be useful as jumping off points for developing *testable* theories."

"Oh, it's okay, I took no offense."

🧍 "Actually, we already have a theory that explains everything I've demonstrated," continues our physicist, "and vastly more. It's quantum theory. It's basic to all of physics and chemistry, and much of modern technology. Even theories of the Big Bang are based on quantum theory."

"Why didn't you use it to explain your demonstrations?" questions a woman sitting with her chin in her hands.

"I might have done that," replies our physicist, "but I wanted to make an important point: The remarkable thing I've demonstrated—that the physical condition of the marble depends on your free choice of experiment—arises *directly from the experimental facts.*

The enigma is posed by the quantum *experiment*. It is not merely *theoretical*. But now that you've seen the demonstration, let me tell you quantum theory's explanation of what we've seen."

"My apparatus," she continues, "puts a marble in each box *pair*, but it does not put that marble in a *single* box. Quantum theory tells us that before you looked, the marble was in what we call a 'super-position state' simultaneously in both boxes. Your gaining knowledge of it being in a particular box *caused* it to be wholly in that box. Even if you gained that knowledge by finding one box empty and did not even see the marble at all, your merely gaining the *knowledge* that it was in the other box would cause it to be wholly in that box. Gaining knowledge in any way whatsoever is enough."

The GROPE (being a Group of Rational and Open-minded PEople) listens politely. But what our physicist said is not readily accepted.

A man suddenly blurts out: "Are you claiming that before we looked and found the marble in one of the boxes, it wasn't there, that our looking created the marble there? That'd be silly."

"Wait, I think I understand what she's saying," the woman sitting next to him volunteers. "I've read about quantum mechanics. I think she just means that the marble's wavefunction, which is the probability of where the marble is, was in both boxes. The actual marble was, of course, in one box or the other."

 "The first part of what you said is OK," says our physicist encouragingly. "What was in each of the boxes was indeed half of the marble's wavefunction. The waviness is the probability of finding a marble in the box. But there is no 'actual marble' in addition to the wavefunction of the marble. The wavefunction is the *only* thing that physics describes. It's therefore the only *physical* thing."

Our physicist sees frowns and eyes rolled upward. She is glad they are (supposedly) open-minded. "Watch how nicely quantum theory explains the pattern we get when I open the boxes at the same time," she continues. "The parts of the wavefunction that were in each box both spread out over the detecting screen."

Moving her two hands wavelike as she talks: "The two parts of the wavefunction are waves coming out of each box to the screen.

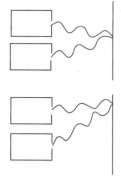

Figure 8.7
Reinforcement and
cancellation of waves
from two boxes

At some places on the screen, wave crests from one box arrive at the same time as crests from the other box, and the wavefunction from the two boxes at that place on the screen add together.

That's a place with large waviness, with a large probability of finding a marble. At other places on the screen, crests from one box come at the same time as troughs from the other box, and the parts of the wavefunction from the two boxes cancel each other. That's a place with zero waviness, with zero probability of finding a marble. That's how each marble follows the rule telling it where it can land. This addition and cancellation of waves is called 'interference' and explains the so-called 'interference pattern' we see."

Satisfied that she has made her point, our physicist stands smiling with her hands on her hips.

A thoughtful-looking woman slowly responds: "I understand how waves, you call them wavefunctions, can create patterns of waviness. We even see that with water waves. But probability has to be the probability *of* something. What's the waviness the probability *of* if it's not the probability of an actual marble being somewhere?"

 "The wavefunction of the marble at some place gives the probability of your *finding* the marble there," our physicist emphasizes. "There was no actual marble there before you looked and found it there."

"I know this isn't easy to accept," she continues sympathetically. "Let me put it in other words. Consider a marble whose wavefunction is equally in both boxes. If you look in either box, you find out where the marble is. The probability then becomes unity in one box and zero in the other. The waviness collapses totally into a single box. That concentrated waviness, which your observation *created*, is what you call the actual marble. But our being able to see an interference pattern proves there was no actual marble in a single box, before you looked."

"Just wait a second!" says a fellow who's been shaking his head for some time. "You put it in other words, so I'll put it in other words: If quantum theory says that by observing something some-place, I create it there, it's saying something ridiculous!"

 "Would you say, 'shocking,' perhaps?" replies our physicist. "Niels Bohr, a founder of quantum theory, once said that anyone not shocked by quantum mechanics has not understood it. But no pre-diction of the theory has ever been wrong. You agree, don't you, that making consistently correct predictions is the *only* criterion a scientific theory need satisfy? That's been the method of science since Galileo. Whether or not it fits our intuitions should be irrelevant."

At this point, another member of the GROPE can no longer contain herself: "If you're saying that unobserved things are just probabilities, that nothing's real until we observe it, you're saying we live in a dream world. You're trying to foist some silly solipsism on us."

 "Well," our physicist replies calmly, "there's a saving grace. The big things we actually deal with are real enough. Remember, you need to do an interference-type experiment to actually *demonstrate* the creation by observation. And it's not practical to do that with big things. Our marbles are very small. So, for all *practical* purposes, there's no need for concern."

While that disturbed member fumes silently, another raises a hesitant hand and says: "If little things are not real, how can big things be real? After all, a big thing is just a collection of little things. A water molecule is just one atom of oxygen and two of hydrogen, and an ice cube is just a collection of water molecules, and a glacier is just a big ice cube. Do we create the glacier by looking at it?"

 Our physicist is now visibly uncomfortable. "Well, in a sense . . ., it's sort of complicated . . . but, as I said, for all practical purposes it doesn't really matter, so. . . ." Then noticing a member of the GROPE with a friendly expression, our physicist invites his comment with a smile.

Trying to be conciliatory, he volunteers: "Maybe what you're driving at is the notion that 'We create our own reality.' I sometimes feel much that way."

"Oh, I can go along with that," our physicist nods. "But that kind of 'reality' is something different. When I say, 'I create my own reality,' I'm talking of *subjective* reality. I'm saying that I accept responsibility for my personal perceptions and my social situation. Something like that at least. The reality we're talking about here is *objective* reality, physical reality. An observation creates an objective situation, which is the same for everyone. After you look into one of the boxes and collapse the wavefunction of the marble wholly into that particular box, anyone else who looks will find the marble there— even though we could have shown that it wasn't wholly there before you looked."

That member of the GROPE who had been fuming in silence now speaks up a bit too loudly: "This reality creation you're talking about is crazy! Your quantum theory may *work* perfectly, but it's absurd! Are people letting you physicists get away with this?"

 "I suppose so," replies our physicist.

"Then you're getting away with murder!"

 "Well, we usually keep the skeleton in the closet."

We trust you could identify with the GROPE. We do, at least when we open mindedly try to understand what's really going on. The best thing to do when you're baffled is to go back and ponder the theory-neutral demonstration of the brute facts, our box-pairs version of the archetypal quantum experiment.

We'll see how to stop worrying and love quantum mechanics, *pragmatically* at least, when we come to the Copenhagen interpretation in chapter 10.

9

One-Third of Our Economy

Developing quantum theory was "the crowning intellectual achievement of the last century," says California Institute of Technology physicist John Preskill. It's the underlying principle for many of today's devices, from lasers to magnetic resonance imaging machines. And these may prove to be just the low-hanging fruit. Many scientists foresee revolutionary technologies based on the truly strange properties of the quantum world.

—*Business Week*, March 15, 2004

We were deep into the quantum mysteries in our "Quantum Enigma" course, which is addressed to students not majoring in the sciences (though some physics majors always take it). A young woman's hand went up with a question: "Is quantum mechanics useful for anything *practical?*" I (Bruce) was speechless for at least ten seconds. In the narrowness of my physicist perspective, I had just assumed that everyone realized that quantum mechanics played a big role in our technology. Putting aside my lecture notes, for the rest of the hour I went off on a tangent to tell of practical applications of quantum mechanics.

This short chapter takes us off on that same tangent. The theme of our book is presenting the undisputed quantum facts that reveal physics' encounter with consciousness. But the same quantum facts are basic to both modern science and today's technology. After flirting with consciousness and free will in our last chapter, it's good to make contact with the solid ground of practical applications before taking off again.

Quantum mechanics is essential to every natural science. When chemists do more than deal with empirical rules, their theories are fundamentally quantum mechanical. Why grass is green, what makes the sun shine, or how quarks behave inside protons are all questions that must be answered quantum mechanically. The still-to-be-understood nature of black holes or the Big Bang is sought in quantum terms. String theories, which may hold a clue to such things, all start with quantum mechanics.

Quantum mechanics is the most accurate theory in all of science. An extreme test is the calculation of the "gyromagnetic ratio of the electron" with a precision of one part in a trillion. (What the gyromagnetic ratio is doesn't matter here.) Measuring something to a part in a trillion is like measuring the distance from a point in New York to a point in San Francisco to better than the thickness of a human hair. The measurement was done, and the theory's prediction was right on the mark.

Although quantum mechanics works well in science, how important is it *practically*? In fact, one-third of our economy involves products based on quantum mechanics. We'll describe four technologies where the quantum aspects are right up front: the laser, the transistor, CCDs (charge-coupled devices), and MRI (magnetic resonance imaging). We won't go into detail—our point is just to show how quantum phenomena enter the picture, and how practical physicists and engineers deal with the contradictory properties of microscopic entities.

The Laser

Lasers come in wide variety. Some are many meters long and weigh tons. Others extend less than a millimeter. The beam of red light scanning bar codes at the supermarket checkout counter comes from a laser. A laser reads DVDs and writes in laser printers. A powerful laser can drill through concrete. Lasers produce the light for fiber optic communication making possible the Internet. They set lines for surveyors and guide "smart bombs." With a sharply focused laser, a surgeon can pin down a detached retina. New applications continually appear in medicine, communications, computation, manufacturing, entertainment, warfare, and fundamental science.

The physics basic to the laser came in 1917 (a decade before the Schrödinger equation). Einstein predicted that photons impinging on atoms in an excited state would stimulate the emission of identical further photons. Almost forty years later, Charles Townes, seeking a way to generate very short wavelength microwaves, realized a solution in the phenomenon of stimulated emission. The first application used molecules of ammonia in an excited state to amplify microwaves.

Townes, of course, realized that the physical process involved was applicable at all frequencies, in particular, at light frequencies. He suggested the acronym "laser" for Light Amplification by the Stimulated Emission of Radiation. (The original microwave device was called a "maser.") Only a few years later, laser action was demonstrated by flashing an intense light on a crystal of synthetic ruby, exciting the chromium atoms in the crystal to a state from which they then radiated identical photons by stimulated emission. The report of that surprisingly quick result was actually rejected as likely mistaken by the prestigious American physics journal to which it was submitted. But it was soon published by the British journal *Nature*.

A laser produces a narrow beam of light of a single frequency that can be focused down to a tiny spot. Inside one type of laser, a photon of the proper frequency, hitting an atom in an excited state, stimulates the emission of a second photon of exactly the same frequency traveling in exactly the same direction—a clone. Where we had one photon, we now have two. If we maintain many atoms in the excited state, this process can continue in a chain reaction to produce many identical photons.

A problem the laser designer must surmount is that the likelihood of a photon hitting an atom on a single pass through the lasing material is small. The light is therefore reflected back and forth through the lasing material between a pair of mirrors. A resonating guitar string must vibrate in an integral number of half wavelengths. Likewise, laser mirrors must be spaced by an integral number of half wavelengths of the light. One of the mirrors is slightly transparent, allowing a bit of the beam to exit the laser on each bounce.

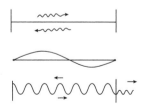

Figure 9.1 Top: Light waves reflected back and forth between two mirrors. Center: A resonating guitar string. Bottom: A light wave resonant between two laser mirrors

Notice how we slipped from talking of light being a stream of compact photons, each one hitting a single atom, to light being an extended wave stretching between two macroscopic mirrors. This is analogous to our atom that could be compactly concentrated in a single box or be a wave spread over two boxes. Laser designers must think in both contradictory ways, but not at the same time.

The Transistor

The transistor is surely the most important invention of the twentieth century. Modern technology would be impossible without it. A transistor controls the flow of an electric current. Before the transistor was developed in the 1950s, such operations were done by vacuum tubes. Each tube was as large as your fist, gave off almost as much heat as a light bulb, and cost several dollars.

Today, billions of transistors on a single chip cost less than a millionth of a cent each, and each is only millionths of an inch across. A personal computer has billions of them. Using vacuum tubes, a computer with the power of a modern laptop would be ridiculously expensive, occupy vast territory, and require the electric power of a major city's generating plant.

Transistors are everywhere: in computers, TVs, cars, phones, microwave ovens, and the watch on your wrist. Modern life continues to change because of the increasing number of transistors that can be put on a single chip. In 1965, Gordon Moore, co-founder of Intel, the world's largest semiconductor chip maker, predicted that the number of transistors that could be put on a single chip would double every eighteen months. "Moore's law" has held up remarkably well over four decades. From thousands of transistors per chip in the 1970s to millions in the 1990s, we now have billions.

In 2009, researchers were able to change the state of a single benzene molecule by applying a voltage to it, and by doing that they could control the current flowing through it. The single benzene molecule behaved like a macroscopic transistor.

We seem to be hitting some fundamental physical limits to how small transistors can be. The end of Moore's law may be in sight, though this has been said before and has been wrong. With the possibility of the quantum computers we discuss below, the limit of what can be *accomplished* on a chip may still be beyond our estimating.

How do transistors involve quantum phenomena? Most transistors are based on silicon, each atom of which has fourteen electrons. Of these, ten electrons are held to their parent nucleus. The other four are "valence electrons" that bind each silicon atom to its neighbors. Valence electrons are not held to a parent nucleus. Each one extends throughout the silicon crystal. Each valence electron is simultaneously *everyplace* in the crystal.

The electrons directly involved in current flow in the transistor are another matter. These are released by different atoms, phosphorus, for example, which are added to the silicon crystal. Designers of transistors must concern themselves with these released "conduction electrons" being slowed by bumping into individual impurity atoms or being trapped by such impurities. They must consider conduction electrons as objects that are compact on the atomic scale.

How do the engineers and physicists who design lasers and transistors deal with photons and electrons that are sometimes compact, smaller than atoms, and that sometimes extend over macroscopic distances? They cultivate a benign schizophrenia. They just learn when to think one way and when to think the other way. And, for all practical purposes, that's good enough.

Charge-Coupled Devices (CCDs)

Physics Nobel Prizes are usually awarded for fundamental discoveries far from any practical application. In 2009, however, two major technological accomplishments were honored: fiber optics and charge-coupled devices (CCDs). They have each had great impact in both science and our economy.

CCDs, converting incoming light into an electrical signal, have taken over and greatly expanded personal photography, have revolutionized astronomy, and are steadily improving diagnostic medicine. A typical digital camera has a semiconductor chip with many millions of CCDs.

A CCD used optically is integrated with a photoelectric element. The phenomenon that inspired Einstein to postulate the photon in 1905 starts the process in a CCD. In the original photoelectric effect, photons eject electrons from the surface of a metal into the vacuum, where they can then be controlled by an electric field. In the CCD, photons excite a cluster of electrons in silicon to states in which they can be moved by an electric field.

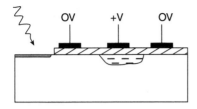

Figure 9.2 Schematic diagram of a CCD with a photoactive region at the left. A group of electrons being transported is shown beneath the center electrode

A nearby metal electrode is given a positive charge to attract that electron cluster. Subsequently, that positive charge is turned off and a neighboring electrode is charged positively to attract that cluster (to the right in Figure 9.2). This process is repeated by a clocked signal to bring the electron cluster to transistors that record the charge. The arrival time of a particular cluster gives the position of particular photons in the image.

CCDs can detect even a single photon, giving CCDs much greater light sensitivity than photographic film. Moreover, since the number of electrons excited is directly proportional to the number of incident photons, great image accuracy is possible. Moreover, with CCDs an image or other data is presented in digital form available for processing or analysis.

Magnetic Resonance Imaging (MRI)

Magnetic resonance imaging (MRI) produces strikingly clear and detailed images of any desired tissue in the body. It is on the way to becoming medicine's most important diagnostic tool. Presently, most MRI machines are large and expensive, costing more than a million dollars, and an MRI examination can cost more than a thousand. Fortunately, size and costs are coming down even as diagnostic capabilities increase.

Magnetic resonance images determine the distribution of a given element, usually hydrogen, in a particular material in the region of the body examined. Different tissues, bone or flesh, tumor or normal, are pictured by the differing concentrations of a particular chemical substance.

The details of MRI are complicated, but the only point we wish to make is that physicists and engineers developing MRI must explicitly use quantum mechanics. The basic idea is the magnetic resonance of atomic nuclei. (Magnetic resonance imaging was originally called NMRI, "*nuclear* magnetic resonance imaging," before the anxiety-causing n-word was dropped.)

Nuclei are little magnets having a north and a south pole. In a magnetic field, the hydrogen nucleus, which is a proton, is "spatially quantized."

That is, it has two states: In one, its north pole points up along the magnetic field; in the other, its north pole points down against the field. In an MRI machine, an electromagnetic wave of the proper frequency puts the hydrogen nuclei in the particular site in the body being imaged at that moment into a quantum superposition state in which their north poles point up and down simultaneously. These nuclei radiate electromagnetic waves as they return to their lower state, and the amount of this radiation reveals their concentration in a particular region. It then requires extensive computation to produce an image from this data.

Crucial to making MRI practical is the vast increase in computational power made possible by the increasing number of transistors on a single chip. When the basic of idea of MRI was proposed, its significance was not recognized, probably because the required computational power was not envisioned. The paper proposing it was thus initially rejected by the journal to which it was submitted.

Most MRI machines involve a several-ton superconducting magnet held at a temperature only a few degrees above absolute zero. In a superconducting metal, electrons condense into a quantum state in which they move as a bound-together unit, each electron simultaneously everyplace in a large mass of metal. It requires a substantial quantum of energy to remove an electron from this moving, bound-together unit. Therefore, once the superconducting electrons are given an initial push, no electric power is needed to maintain the electron current flow and the magnetic field.

MRI is made possible by the coming together of the quantum phenomena responsible for nuclear magnetic resonance, superconductivity, and the transistor. Each of these technologies, as well as the laser and the CCD, has led to a Nobel Prize in Physics, most recently MRI in 2004 and the CCD in 2009.

The Future

Quantum Dots

The involvement of quantum phenomena in technology, and even biotechnology, expands rapidly. In 2003, the journal *Science* named "quantum dot" research as one of the top scientific breakthroughs of the year. Quantum dots, each made of a few hundred or fewer atoms, are artificial constructs with the quantum properties of a single atom, a series of discrete energy levels, for example. Quantum dots have been designed to reveal the workings of the nervous system, to be ultra-sensitive detectors of breast cancer, or produce versatile pigments. When electrodes are attached to quantum dots, they can be used to control current flow as ultrafast transistors or to process optical signals. In 2009, researchers at Canada's National Institute for Nanotechnology produced single-atom quantum dots where one dot could control electron motion in a neighboring pair of dots. Since this could be done at room temperature, practical use may not be very far off. In 2010, work with quantum dots indicates their use could increase the efficiency of solar cells to more than 60% from the current theoretical limit of 30%. Expect to hear a lot about quantum dots in the future.

Quantum Computers

An operating element in a classical digital computer must be in one of two states: either "0" or "1." An "unobserved" operating element in a *quantum* computer can be in a superposition state simultaneously "0" *and* "1." This is much like the situation we described where a single unobserved atom was in a superposition state simultaneously in each of two boxes.

While each element in a classical computer can deal with only a single computation at a time, superposition allows each element in a quantum computer to deal with many computations simultaneously. This vast parallelism would enable a quantum computer to solve in minutes certain problems that would take a classical computer a billion years. Early on, it was thought that the kind of computation quantum computers could do much faster than classical computers was very limited, but those limits are expanding. In particular, quantum computers should excel in the search of large databases.

Commercial applications are pursued, and promising results are constantly reported. But you won't be able to buy a quantum laptop anytime soon. Quantum computers face serious technical problems. As with an atom in a pair of boxes, the wavefunctions of the logic units of a quantum computer are exceedingly fragile. When objects interact, their wavefunctions become "entangled," and entanglement is basic to the operation of a quantum computer. The computer logic units must be properly isolated from the random thermal environment, which would otherwise rapidly disrupt the *intended* entanglements. Encouragingly, encoding techniques increasing by a factor of one hundred the time quantum states can tolerate such disruption have recently been developed. IBM, taking quantum computing seriously, has recently assembled a large research group to begin a five-year project.

Engineers and physicists who work with modern technologies may deal with quantum mechanics on an everyday basis, but they need not face up to the issues raised by the quantum mysteries. Many are not even aware of them. In teaching quantum mechanics, physicists, including ourselves, spend little, if any, time on the enigmatic aspects. We concentrate on the practical stuff students will need to use. Might we also avoid the enigma because our "skeleton in the closet" can be a bit embarrassing? Our next chapter on the Copenhagen interpretation of quantum mechanics presents our discipline's standard way of putting aside the enigma, for all practical purposes at least.

10

Wonderful, Wonderful Copenhagen

Wonderful, wonderful Copenhagen . . .
Salty old queen of the sea
Once I sailed away
But I'm home today
Singing Copenhagen, wonderful, wonderful
Copenhagen for me.

—"Wonderful Copenhagen," by Frank Loesser

The meaning of Newton's mechanics was clear: It described a reasonable world, a "clockwork universe." Classical physics needed no interpretation. Einstein's relativity is surely counterintuitive, but we need no "interpretations" of relativity. After working with relativity for a while, we readily accept the idea that moving clocks run slow. The quantum theory assertion, that observation *creates* the reality observed, is harder to accept—it needs interpretation.

Students come into physics to study the physical world. The *Oxford English Dictionary* defines this sense of "physical" well: "Of or pertaining to material nature, *as opposed to the psychical, mental, or spiritual*" (emphasis added). The *New York Times* in 2002 quoted science historian Jed Buchwald: "Physicists . . . have long had a special loathing for admitting questions with the slightest emotional content into their professional work." Indeed, most physicists want to avoid dealing with the skeleton in our closet: physics' encounter with the conscious observer. The Copenhagen interpretation of quantum mechanics allows that avoidance. It's been called our discipline's "orthodox" position.

The Copenhagen Interpretation

Niels Bohr recognized early on that physics had encountered the observer and that the issue had to be addressed:

> The discovery of the quantum of action shows us, in fact, not only the natural limitation of classical physics, but by throwing a new light upon the old philosophical problem of the objective existence of phenomena independently of our observations, confronts us with a situation hitherto unknown in natural science.

Within a year after Schrödinger's equation, the Copenhagen interpretation was developed at Bohr's institute in Copenhagen, with Bohr as its principal architect. Werner Heisenberg, his younger colleague, was the other major contributor. There is no "official" Copenhagen interpretation. But every version grabs the bull by the horns and asserts that an observation *produces* the property observed. The tricky word here is "observation." Does "observation" require *conscious* observation? It depends on the context. (When we specifically mean *conscious* observation, we'll try to say so.)

Copenhagen broadens the assertion that observation produces the properties observed by defining observation as taking place whenever a microscopic, an atomic-scale object, interacts with a macroscopic, large-scale object. When a piece of photographic film is hit by a photon and thus records where the photon landed, the film "observes" the photon. When a Geiger counter clicks as an electron enters its discharge tube, the counter "observes" the electron.

The Copenhagen interpretation thus considers two realms: the macroscopic, classical realm of our measuring instruments governed by classical physics; and the microscopic, quantum realm of atoms and other small things governed by the Schrödinger equation. It argues that since we never deal *directly* with the quantum objects of the microscopic realm, we need not worry about their physical reality, or their lack of it. An "existence" that allows the calculation of their effects on our macroscopic instruments is sufficient. After all, it is only the behavior of our classical apparatus that we report. Since the difference in scale between atoms and Geiger

Figure 10.1 Drawing by Michael Ramus, 1991. © American Institute of Physics

counters is so vast, Copenhagen treats the microscopic and macroscopic realms separately.

We often speak of the behavior of electrons, atoms, and other microscopic objects as if we observed them directly, as if they were as real as little green marbles. (We might, for example, say: "An alpha particle bounced off a gold nucleus.") It is, nevertheless, only the response of our laboratory instruments that we need consider real.

In 1932, just a few years after Bohr's Copenhagen interpretation, John von Neumann presented a rigorous treatment that is also referred to as the Copenhagen interpretation. He showed that if quantum mechanics applies universally, as claimed, an *ultimate* encounter with consciousness is inevitable. However, von Neumann showed that for all *practical* purposes, we may consider macroscopic apparatus classically. This emphasizes that Bohr's separation of the microscopic and the macroscopic is only a very good approximation. We discuss von Neumann's conclusion in chapter 17. It warns that whenever we refer to "observation," the question of consciousness lurks.

Most physicists, wishing to avoid philosophical problems, readily accept Bohr's Copenhagen interpretation. We will later see physicists occasionally sail away to speculative shores, but when we actually *do* physics or teach physics, we all come home to wonderful Copenhagen.

Physicists today show more unease with atoms being less than real as technology invades the ill-defined territory between the classical and the quantum realms. We will therefore critically explore the Copenhagen interpretation, the tacitly accepted stance of working physicists.

What Copenhagen Must Make Acceptable

While we presented physics' "skeleton in the closet" in chapter 8 as a story, real versions of such experiments are done all the time. We display those

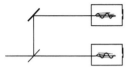

contradictory results (using photons or electrons) even as lecture demonstrations.

In that chapter 8 story, a small object is sent into a pair of well-separated boxes. Looking into the boxes,

Figure 10.2 The atoms-in-a-box-pair demonstration

you always find the whole object in a *single* box, and the other box empty. According to quantum theory, however, before the object was observed, it was simultaneously in *both* boxes, *not* wholly in a single box. An interference experiment, which you could have chosen, would have established that. By your free choice, you could establish either of two contradictory prior realities. Even though today's technology limits the display of quantum phenomena to very small things, quantum theory presumably applies to everything—to baseballs as well as atoms. Copenhagen must make this weirdness acceptable.

Three Pillars of Copenhagen

The Copenhagen interpretation rests on three basic ideas: The probability interpretation of the wavefunction, the Heisenberg uncertainty principle, and complementarity. We look at these in turn.

The Probability Interpretation of the Wavefunction

We've been using the idea all along that the waviness in a region (technically, the absolute square of the wavefunction) is the probability that the object will be *found* in that region. This probability interpretation of waviness is central to the Copenhagen interpretation.

While classical physics is strictly deterministic, quantum mechanics displays Nature's intrinsic randomness. On the atomic level, God plays dice, Einstein notwithstanding. (Einstein repeatedly emphasized that observer-created reality, not randomness, was his real problem with quantum theory.) That Nature is ultimately probabilistic is not hard to accept. After all, much of what happens in everyday life has randomness. Were that the whole story, there would be much less concern with a "quantum enigma." Probability in quantum mechanics implies more than randomness.

Classical probability, in the shell game say, is the *subjective* probability (for you) of where the pea is. But there is also a real pea under one shell or the other. Quantum probability is *not* the probability of where the atom *is*. It's the objective probability of where you, or anybody, will *find* it. The atom wasn't someplace until it was *observed* to be there.

Since quantum theory has no atom in addition to the wavefunction of the atom, if an atom's wavefunction occupies two boxes, the atom itself is simultaneously in both boxes. Its later observation in a single box *causes* it to be *wholly* in that box (or to be *not* in that box).

The point of that last paragraph is hard to accept. That's why we keep repeating it. Even students completing a course in quantum mechanics, when asked what the wavefunction tells, often incorrectly respond that it gives the probability of the object *being* at a particular place. Wrong. The wavefunction gives the probability of *finding* the object at a particular place. The senior-level text we teach from (by Griffiths, and listed in "Suggested Reading") emphasizes the correct point by quoting Pascual Jordan, one of the founders of quantum theory: "Observations not only *disturb* what is to be measured, they *produce* it." But we're sympathetic with our students. Calculating wavefunctions is hard enough without worrying about their deeper meaning.

Though we've been speaking of "observation," we've not really said what constitutes an observation. It's ultimately a controversial issue. But there are clear-cut cases.

For a photon bouncing off an atom, there is a clear answer: The photon does *not* observe the atom. After the encounter, the photon and atom are together in a superposition state that includes all possible positions of the photon and the atom. This can be confirmed with a complex two-body interference experiment. At the other extreme, when we hear the click of the Geiger counter in contact with the rest of the world recording that

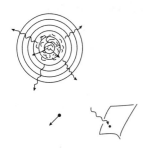

Figure 10.3 Bouncing a photon off an atom does not create the atom's position until the photon is detected

photon, an observation of the photon's, and thus the atom's, position has been made.

It's the in-between situations that are controversial. Strictly speaking, of course, the Geiger counter must obey quantum mechanics. Were it isolated from the rest of the world, it would merely join the superposition state of the microscopic object encountering it. It would thus not "observe." But for practical reasons, it is essentially impossible to demonstrate a large object to be in a superposition state because it is essentially impossible to isolate a large object from the rest of the world. We'll talk more of this difficulty in our next chapter.

Let's be careful about what it means to be "unobserved." Consider our atom in its box pair. Until the position of the atom in a particular box is observed, the atom doesn't exist in a *particular* box. We nevertheless initially "observed" the atom when we grabbed it and sent it into a box pair. The atom's position in the *pair* of boxes is thus an observed reality. However, taking the extreme case of very large boxes, we can say the atom essentially has no position at all. It does not *have* the property of position. The same can be said of any unobserved property of an object.

The Copenhagen interpretation generally adopts the view that only the *observed* properties of microscopic objects exist. John Wheeler put it concisely: "No microscopic property is a property until it is an observed property."

If we carry this to its logical conclusion, microscopic objects themselves are not real things. Here's Heisenberg on this:

> In the experiments about atomic events we have to do with things and facts, the phenomena that are just as real as any phenomena in daily life. *But the atoms or elementary particles themselves are not real*; they form a world of potentialities or possibilities rather than one of things or facts. (emphasis added)

According to this view, atomic-scale objects exist only in some abstract realm, not in the physical world. If so, it's OK that they don't "make sense."

It's enough that they affect our measuring instruments in accord with quantum theory. Big things, on the other hand, are real, for all practical purposes. Of course, their classical description is only an *approximation* to the correct quantum laws of physics. If so, the microscopic realm, the *unobserved* realm, is in some sense the more real. Plato would like that.

If the microscopic realm consists merely of possibilities, how does physics account for the small things that big things are made of? The most famous statement on this is a bold one often attributed to Bohr:

> There is no quantum world. There is only an abstract quantum description. It is wrong to think that the task of physics is to find out how nature *is*. Physics concerns what we can *say* about nature.

This is actually probably a summary of Bohr's thinking by one of his associates. But it fits with what Bohr has said in more complicated ways. The Copenhagen interpretation avoids involving physics with the conscious observer by redefining what has been the goal of science since ancient Greece: to explain the actual world.

Einstein rejected Bohr's attitude as defeatist, saying he came to physics to discover what's really going on, to learn "God's thoughts." Schrödinger rejected the Copenhagen interpretation on the broadest grounds:

> Bohr's standpoint, that a space-time description [where an object *is* at some time] is impossible, I reject at the outset. Physics does not consist only of atomic research, science does not consist only of physics, and life does not consist only of science. The aim of atomic research is to fit our empirical knowledge concerning it into our other thinking. All of this thinking, so far as it concerns the outer world, is active in space and time. If it cannot be fitted into space and time, then it fails in its whole aim, and one does not know what purpose it really serves.

Would Bohr actually deny that a goal of science is to explain the natural world? Perhaps not. He once said: "The opposite of a correct statement is an incorrect statement, but the opposite of a great truth may be another great truth." Bohr's thinking is notoriously hard to pin down.

A colleague of Heisenberg's once suggested that the wave–particle problem is merely semantic and could be solved by calling electrons neither waves nor particles but "wavicles." Heisenberg, insisting that the philosophical issues raised by quantum mechanics included the big as well as the small, replied:

> No, that solution is a bit too simple for me. After all, we are not dealing with a special property of electrons, but with a property of all matter and of all radiation. Whether we take electrons, light quanta, benzol molecules, *or stones*, we shall always come up against these two characteristics, the corpuscular and the undular. (emphasis added)

He's telling us that, in principle (and that is what's important to us here), *everything* is quantum mechanical and ultimately subject to the enigma. This brings us to the second pillar of the Copenhagen interpretation, the uncertainty principle, the idea for which Heisenberg is most widely known.

The Heisenberg Uncertainty Principle

Heisenberg showed that any demonstration to refute the claim of observer-created reality would be frustrated. Here's his example:

While doing an interference experiment, look to see out of which box each atom *actually* came. Seeing the atom to have come from a *single* box would demonstrate that the atom had actually *been* in that single box. If it then followed the interference rule implying that it came out of *both* boxes, quantum theory would be shown inconsistent, therefore wrong. To show that such a demonstration must fail, Heisenberg produced the thought experiment now called the "Heisenberg microscope."

To see out of which box an atom came, you could bounce light off it. Reflecting light from something is our usual way of seeing things. In order not to kick the atom hard enough to deflect it from going only to an allowed place in the interference pattern, hit it as gently as possible. Use the least possible light, a single photon.

In general, if two things are closer together than the wavelength of the waves coming from them, their separation cannot be established. This is illustrated in figure 10.5. In figure 10.5a, the wavelength, the distance between crests, is smaller than the separation of the wave sources. The crests coming to the observing "eye" are clearly coming from two different directions, from different places.

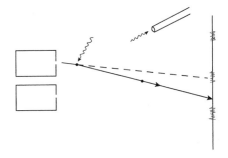

Figure 10.4 The Heisenberg microscope

In figure 10.5b, the wavelength is larger than the separation of the wave sources. The crests coming to the observing "eye" are *not* clearly distinguishable as coming from two different directions, from different places. Therefore, to tell which box the atom came from, the wavelength of the light reflected from the atom must be smaller, shorter, than the separation of the boxes.

But a short wavelength means a large number of crests coming per second. That's a high frequency, and a high-frequency photon has a high energy. It would give the atom a hard kick. Heisenberg easily calculated that a photon with a wavelength short enough to *distinguish* the two sources would kick an atom hard enough to smear any interference pattern. It would cause some atoms go to places forbidden by the interference rule. The dotted line in figure 10.4 shows the path the atom would have taken were it *not* hit by the photon.

The Heisenberg microscope story shows that if you saw each atom come from a particular single box, you could not *also* see an interference pattern showing that each atom had been in both boxes. You therefore could not refute the observer-creation of reality in this case. Heisenberg proudly came to Bohr with his discovery. Bohr was impressed but told his young colleague that he didn't have it quite right. Heisenberg forgot that if you knew the angle at which the photon bounced off, you could in fact *calculate* which box the atom came from. But he had the right basic idea. Bohr showed him that by including the size of the microscope lens needed to measure the photon

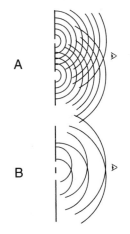

Figure 10.5 a. Sources of waves separated by more than a wavelength. b. Sources of waves separated by less than a wavelength

angle in his analysis, he could recapture the result he thought he had. Missing this point doubly embarrassed Heisenberg. He reports that determining the direction of a light wave with a microscope was a question he had missed on his doctoral exam.

Heisenberg went on to generalize his microscope story to become the "Heisenberg uncertainty principle": The more accurately you measure an object's position, the more uncertain you will be about its speed. And vice versa, the more accurately you measure an object's speed, the more uncertain you will be about its position.

The uncertainty principle can also be derived directly from the Schrödinger equation. In fact, the observation of any property makes a "complementary" quantity uncertain. Position and speed (or momentum) are, for example, complementary quantities. Energy and the time of observation are another complementary pair. The bottom line is that the observation of any property disturbs things enough to prevent refutation of quantum theory's assertion that observation *creates* the property observed.

We note in passing that the uncertainty principle arises in some discussions of free will. In the worldview of classical physics, if an "all-seeing eye" knew the position and velocity of every object in the universe at one moment, the entire future could be predicted with certainty. To the extent that we are part of this physical universe, classical physics rules out free will. Because the uncertainty principle denies this Newtonian determinism, it has entered philosophical discussions of determinism and free will. Uncertainty can *allow* free will by denying determinism, but randomness, quantum or otherwise, is not free choice. Quantum uncertainty cannot *establish* free will.

Complementarity

While the uncertainty principle shows that any observation of an object necessarily disturbs that object enough to preclude the falsification of quantum theory, that's not enough. We also need the Copenhagen interpretation's third pillar: complementarity. It's the hard one to accept. (Complementarity is what truly bothered Einstein, not the randomness of his dice quip.)

Consider a set of 1,000 box pairs, each pair containing an atom in a superposition state simultaneously in both boxes. Look in one box of every pair. About half the time you will see an atom in the box you opened. According to the uncertainty principle, seeing that atom disturbed it with the photons you shined in. Therefore, toss away all the box pairs for which you saw, and thus disturbed, an atom. You are left with about 500 box pairs whose atoms were *not* physically disturbed; no photons bounced off them, the box you looked in was completely empty. But for these box pairs, you *know* which box each atom is in: the box you did *not* look in.

Attempt an interference experiment with these 500 box pairs. An interference pattern would prove that each of those atoms had been simultaneously in *both* boxes of its pair. But for these box pairs, you had already determined that each atom was wholly in a *single* box. Finding an interference pattern would therefore show an inconsistency in quantum theory.

In fact, these supposedly undisturbed atoms do *not* produce an interference pattern. What caused these undisturbed atoms to adopt a different behavior? After all, if you had done an interference experiment before you looked into the empty boxes, these same atoms would have produced an interference pattern.

Although these atoms were not physically disturbed, you did determine which box each atom was in. Apparently, your acquisition of that *knowledge* was sufficient to concentrate each atom totally within a single box. To avoid seeing this as somehow mysterious requires some talk.

The talk we offer in a quantum mechanics class for physics students is that when we look in a box and find no atom, we instantaneously collapse the atom's waviness into the other box. In the shell game, our look collapses the probability, which had been 1/2 for the pea being under each shell, to being zero under the shell we found empty and to 1, a certainty, for the pea being under the other shell. The same thing happens with waviness. After all, waviness is probability.

That explanation is a bit glib. Classical probability *starts out* as merely a measure of one's knowledge. On the other hand, quantum probability, waviness, is supposedly all there is to the physical atom. How did merely acquiring knowledge concentrate the atom in a single box? But we rarely emphasize philosophical conundrums to physics students, who must mainly learn to calculate.

Bohr realized that he had to confront the influence of knowledge on physical phenomena in order to allow physicists to get on with doing physics without getting bogged down in philosophy. He thus asserted his principle of complementarity: The two aspects of a microscopic object, its particle aspect and its wave aspect, are "complementary." A *complete* description requires both contradictory aspects. But we must *consider only one aspect at a time* by specifying the kind of observation we are making, the experiment we are doing.

Accordingly, we avoid the seeming contradiction by considering the microscopic system, the atom, not to exist in and of itself. We must always include in our discussion, implicitly at least, the macroscopic experimental apparatus used to display either one or the other complementary aspect. All is then fine, because it is ultimately only the classical behavior of such apparatus that we report. In Bohr's words:

> The decisive point is to recognize that the description of the experimental arrangement and the recording of observations must be given in plain language, suitably refined by the usual physical terminology. This is a simple logical demand, since by the word "experiment" we can only mean a procedure regarding which we are able to communicate to others what we have done and what we have learnt.
>
> In actual experimental arrangements, the fulfillment of such requirements is secured by the use, as measuring instruments, of rigid bodies sufficiently heavy to allow a completely classical account of their relative positions and velocities.

In other words, although physicists talk of atoms and other microscopic entities as if they were actual physical things, microscopic things are only concepts we use to describe the behavior of our measuring instruments. We need not go beyond describing that behavior in dealing with the microscopic world.

This stance recalls Newton's "*Hypotheses non fingo*" (I make no hypotheses), his claim that an explanation of gravity need not go beyond his equations predicting the motions of the planets. Einstein, of course, with general relativity, his theory of gravity, gave great insight into the nature of space and time by going beyond Newton's equations.

Here's a related tack the flexible Copenhagen interpretation can take: Don't think about experiments that you *might* have done but did not in *fact* do. After all, it was just the perception that we *could* have chosen to do an experiment other than the one we actually did that gave rise to the quantum measurement problem, the quantum enigma.

One eminent physicist emphasized this stance by proclaiming: "Not-done experiments have no results!" Since physics is just about experimental *results*, we need not even think about *not*-done experiments. You couldn't actually *display* a logical contradiction. (A which-box experiment disturbs *those* atoms enough to preclude an interference experiment with them, and after an interference experiment, you can't do a which-box experiment with those *same* atoms.)

Our usual assumption that we could have done other than we actually did is called "counterfactual definiteness." Believing that *if* you did not eat lunch you'd be hungry assumes counterfactual definiteness. We put someone in jail because they *could* have decided to not pull that trigger. We run our lives and society assuming the counterfactual definiteness that this version of Copenhagen denies. Some mathematically inclined physicists do argue that quantum mechanics forces us to just accept a denial of counterfactual definiteness. That can be fine if we just apply quantum theory to microscopic things and don't concern ourselves with its undeniable broader implications.

Denying counterfactual definiteness, the Copenhagen interpretation would seem to deny free will. Is free will an illusion? We can't prove that we're not just automatons in a totally deterministic world that conspires to fool us into believing we make choices. However, we (Fred and Bruce) are each completely confident of our *own* free will, even though neither of us can be *absolutely* sure that his co-author is not a sophisticated robot.

The Acceptance of and the
Unease with Copenhagen

The Copenhagen interpretation asks us to accept quantum mechanics pragmatically. (Bumper-sticker summary of pragmatism: "If it works, it's true.") When physicists want to avoid dealing with philosophy, and for most of us that's almost all the time, we tacitly accept the Copenhagen interpretation. Physicists tend to be pragmatists.

Though, strictly speaking, the properties of microscopic objects are merely *inferred* from the behavior of our apparatus, physicists nevertheless talk of microscopic objects, visualize them, and calculate with models of them as if they were real. But when confronted with paradox, we can always retreat to Copenhagen: The quantum theory of microscopic objects should explain the sensible behavior of our macroscopic equipment, but microscopic objects themselves need not "make sense."

Consider an analogy from psychology (as Bohr did). We report on a person's *behavior*. The *physical* behavior itself presents no paradox. The person's physical movements make sense in that they accord with Newton's law of motion. A person's *motives*, however, are *theories* that should explain the person's behavior. But the motives themselves need not, and often do not, make sense. We pragmatically accept this stance in dealing with people. The Copenhagen interpretation asks us to accept this stance in dealing with microscopic physical phenomena.

If you're not at ease with Copenhagen's solution to the observer problem, you're not alone. We don't know of anyone who understands, and takes seriously, what quantum mechanics is telling us who doesn't admit some bafflement.

Nevertheless, until recently, most quantum mechanics textbooks implied that Copenhagen resolved all problems. One 1980 text dismissed the enigma with a joke, a sketch of a duckbilled platypus labeled "The classical analog of the electron." The idea was that going to the realm of the small, you should be no more surprised by an object being both an extended wave and a compact particle than zoologists going to Australia were surprised by an animal being both a mammal and an egg-laying "duck." In his preface, another 1980s author promises to "make quantum mechanics less mysterious for the student." He does it by never displaying the mystery.

Such attitudes likely stimulated Murray Gell-Mann's remark in accepting the 1969 Nobel Prize in Physics that Bohr has brainwashed generations of physicists into believing the problem had been solved. Gell-Mann's concern is a bit less relevant today since most current quantum texts at least hint of unresolved issues.

Essential to the Copenhagen interpretation was a clear separation of the quantum microworld from the classical macroworld. That separation depended on the vast difference in scale between atoms and the big things we deal with directly. In Bohr's day, there was a wide no-man's land in between. It seemed acceptable to think of the macro realm obeying classical physics and the micro realm obeying quantum physics.

But today's technology has invaded the no-man's land. With appropriate laser light we can see individual atoms with the naked eye the way we see dust motes in a sunbeam. With the scanning tunneling microscope not only can we see individual atoms, but we can pick them up and put them down. Physicists have spelled out their company's name by positioning thirty-five argon atoms.

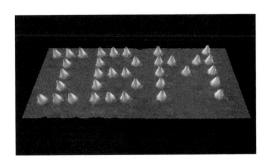

Figure 10.6 Thirty-five argon atoms.
Courtesy IBM

Quantum mechanics is increasingly applied to larger and larger objects. We later discuss the quantum phenomena recently displayed with almost macroscopic—actually visible—objects. Cosmologists write a wavefunction for the whole universe to study the Big Bang. It gets harder today to nonchalantly accept the realm in which the quantum rules apply as not being physically real.

Nevertheless, many physicists, when pressed to respond to the strangeness of the microworld, might say something like: "That's just the way Nature *is*. Reality is just not what we'd intuitively think it to be. Quantum mechanics forces us to abandon naive realism." And they'd leave it at that. Everyone is willing to abandon *naive* realism. But few physicists are willing to abandon "*scientific* realism," defined as "the thesis that the objects of scientific knowledge exist and act independently of the knowledge of them." Quantum mechanics challenges scientific realism.

While few physicists deny quantum strangeness, most probably consider the Copenhagen interpretation, or its modern extension, "decoherence" (discussed in chapter 15), to have taken care of it. For all *practical* purposes, that's all that counts. But more physicists, especially younger physicists, are increasingly open-minded to ideas beyond Copenhagen. Interpretations challenging Copenhagen proliferate. In chapter 15 we discuss several of them. Concern with consciousness itself (as well as its connection with quantum mechanics) has increasingly emerged among physicists, philosophers and psychologists.

The Copenhagen interpretation was recently summarized as "Shut up and calculate!" That's blunt, but not completely unfair. It's the right injunction for most physicists most of the time. The Copenhagen interpretation is surely the wonderful way to deal with quantum mechanics for all *practical* purposes. It assures us that in our labs or at our desks we can use quantum mechanics without worrying about what's "really" going on.

We might, however, wish for more than an algorithm for computing probabilities. Classical physics provided more; it imparted a worldview that changed our culture. It is, of course, a worldview we now know is fundamentally flawed. Can it be that out there in our future there is a quantum impact on our worldview?

A Copenhagen Summary

☏ = Objector

☏ = Copenhagenist

☏ Quantum mechanics violates common *sense*. There must be something *wrong* with it!

☏ No. Never a wrong prediction. It works perfectly.

☏ The better it works, the sillier it looks! It's not logically consistent.

☏ Oh, you know that Einstein *tried* to show that. He gave up.

☏ But quantum mechanics says that little things have no properties of their own, that I actually create what I see by my *looking*.

☏ True. You perceive the basic idea quite clearly.

☏ But with only observer-created properties, little things have no physical *reality*. They're real only when they're being observed. That makes no sense!

☏ Don't worry about "reality," or about "making sense." Small things are only models. Models need not make sense. Models only have to work. Large things are real enough. So everything's just fine.

But a large thing is just a *collection* of little things, of atoms. To be consistent, quantum mechanics would have to say that *nothing* has a reality until it's observed.

Oh, true, if you insist. But it doesn't matter.

Not matter?! If quantum mechanics says my cat and my table aren't real until they're *looked* at, it's saying something *crazy*.

No, it's all ok. You never actually *see* any craziness with big things. For all practical purposes, big things are *always* being looked at.

For all *practical* purposes, sure. But what's the *meaning* of this observer-created reality?

Science provides no meanings. Science just tells us what will happen. It just predicts what will be observed.

I want more than a recipe for making predictions. If you say common sense is wrong, I want to know what's right.

But we've agreed that quantum mechanics is right. The Schrödinger equation tells everything that will happen, everything that can be observed.

I want to know what's really going on, I want the whole story!

The quantum mechanical description *is* the whole story. There's nothing else to tell.

Damn it! There's a real world out there. I want to know the *truth* about Nature.

Science can reveal no real world beyond what is observed. Anything else is merely philosophy. *That's* the "truth"—if you must have one.

That's defeatist! I'll never be satisfied with such a superficial answer. You have science abandoning its basic philosophical goal, its mission to explain the physical world.

Too bad. But don't bother me with philosophy. I've got scientific *work* to do.

Quantum mechanics is manifestly absurd! I won't accept it as a final answer.

(SHE'S NO LONGER LISTENING.)

11

Schrödinger's Controversial Cat

The entire system would [contain] equal parts of living and dead cat.

—Erwin Schrödinger

When I hear about Schrödinger's cat, I reach for my gun.

—Stephen Hawking

By 1935, the basic form of quantum theory was clear. Schrödinger's equation was the new universal equation of motion. Although *required* only for objects on the atomic scale, quantum theory presumably governed the behavior of *everything*. The earlier physics, by then called "classical," was the easier-to-use approximation for big things.

We tell the story Schrödinger invented to show that quantum theory is not only weird, but it's absurd. However, the theory works so well, that most physicists overlook the absurdity. Nevertheless, Schrödinger's story resonates loudly today.

In what follows, when we refer to "quantum theory," we intend its Copenhagen interpretation unless we say otherwise. In this regard, Heisenberg tells us that microscopic objects such as atoms are not "real"— they're just "potentialities." What about things *made* of atoms? Chairs, for example? Is a not-yet-seen galaxy not really there? Pressing such questions, we confront the skeleton physics usually keeps in the closet.

Is it that quantum theory doesn't apply to big things? No, quantum theory underlies *all* physics. We need quantum theory to deal with the underlying basis of large-scale objects, such as lasers, silicon chips, or stars. Ultimately, the working of everything is quantum mechanical. But we don't *see* the quantum weirdness with big things. With the Copenhagen interpretation, Bohr explained that we should apply quantum theory to the little things and deal with big things in classical terms. Most physicists pragmatically accept this injunction and are not bothered by the "non-reality" of the small.

Schrödinger, however, *was* bothered: If quantum theory could deny the reality of atoms, it logically denied the reality of things made of atoms. He felt sure that something this crazy could not be Nature's universal law. We can imagine a conversation between a bothered Schrödinger and a pragmatic young colleague:

SCHRÖDINGER: The Copenhagen interpretation is a cop-out. Nature is trying to tell us something, and Copenhagen is telling us not to listen. Quantum theory is absurd!

COLLEAGUE: But sir, your theory works perfectly. No prediction has ever been wrong. So everything's OK.

S: Come now, I look and I find an atom someplace. The theory says that just before I looked, it wasn't there—it didn't exist at that place. It didn't exist at *any* particular place?!

C: That's right. Before you looked to see where it was, it was a wave-function, just probability. The atom *didn't* exist at any particular place.

S: You're saying my looking *created* the atom to be at the place I found it?

C: Well, yes, sir. That's what your theory says.

S: That's silly solipsism. You're denying the existence of a physically real world. This chair I'm sitting on is a very real chair.

C: Oh, of course, Professor Schrödinger, your chair's real. Only the properties of small things are created by observation.

S: You're saying quantum theory applies only to small things?

C: No, sir, your equation applies to everything. But it's impossible to do an interference experiment with big things. So for all practical purposes there's no reason to worry about the reality of big things.

S: A big thing is just a collection of atoms. If an atom doesn't have physical reality, a collection of them can't be real. If quantum theory says that the real world is created by our looking at it, the theory's absurd!

By a logical technique called *reductio ad absurdum*, or reduction to absurdity, Schrödinger told a story to argue that quantum theory led to an absurd conclusion. Decide for yourself whether to accept his argument. But wait until we also present the standard counter to his reasoning.

The Cat-in-the-Box Story

Our box-pair example is the first step in presenting Schrödinger's argument. Recall that an atom split at a semi-transparent mirror will end up with half its waviness in each of two separate boxes. According to quantum theory, the atom does not exist in one particular box before you *find* the whole atom in one of the boxes. The atom is in a superposition state simultaneously in both boxes. Upon your looking into one box, the superposition-state waviness collapses into a single box. You will randomly find either a *whole* atom in that one box *or* no atom in that box. (You can't choose which!) If you find the one box empty, the atom will be found in the other box. However, with a set of box pairs, you *could* have produced an interference pattern demonstrating that *before* you looked, the atom had been simultaneously in each box.

Our version of Schrödinger's story takes off from here. Suppose that, before we send in the atom, one of the boxes of the pair is not empty. It contains a Geiger counter that will "fire" if an atom enters its box. In firing, this Geiger counter moves a lever to pull the cork from a bottle of hydrogen cyanide. There's also a cat in the box. The cat will die if the poisonous cyanide escapes its bottle. The entire contents of the boxes, the atom, the Geiger counter, the cyanide, and the cat, is isolated and unobserved.

We immediately note that Schrödinger never contemplated actually endangering a cat. This was a *thought* experiment. He referred to the apparatus as a "hellish contraption."

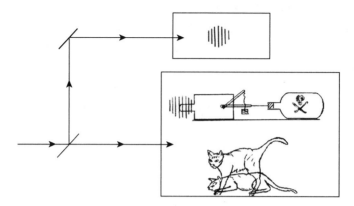

Figure 11.1 Schrödinger's cat

Now, Schrödinger argued, a Geiger counter is just a bunch of ordinary atoms, albeit a complex and well-organized collection. Strictly speaking, it is governed by the same laws of physics that govern the atoms it's made of. It's governed by quantum mechanics. Presumably the same is true for the cat.

Since the waviness of the atom split equally at the semitransparent mirror, half its waviness went into the box with the Geiger counter and the cat, and half went into the other box. As long as the system is not observed in any manner, isolated from the rest of the world, the atom is in a superposition state that we can describe as its being in the box with the Geiger counter and, simultaneously, in the empty box. To be succinct, we say that the atom is simultaneously in both boxes.

The unobserved Geiger counter, which fires if the atom enters its box, must therefore also be in a superposition state. It is both fired and, simultaneously, unfired. The cork on the cyanide bottle is thus both pulled and not pulled. The cat must be both dead and alive. This is, of course, hard to imagine. Impossible to imagine, perhaps. But it's the logical extension of what quantum theory is telling us.

We show quantum theory's version of our yet-unobserved cat, and the rest of Schrödinger's "hellish contraption," with a mixed-metaphor image. We represent the atom by showing the crests of its wavefunction in both boxes. Since the wavefunctions of Geiger counters and cats are too complicated to display, we just picture the Geiger counter both fired and unfired (lever both up and down), the cyanide cork both pulled and not pulled, and the cat simultaneously dead and alive.

What do you see if you now look into the box to see whether the cat is dead or alive? Back when only an atom was in a superposition state in our box pair, any look into a box collapsed the atom totally into one box or the other. Here, a look collapses the wavefunction of the entire system.

Quantum theory predicts a self-consistent situation. If you find the cat dead, the Geiger counter will have fired, the cork on the cyanide bottle will be pulled, and the atom will be in the box with the cat. If you find the cat alive, the Geiger counter will not have fired, the cyanide bottle will be corked, and the atom will be in the other box.

But, according to quantum theory, *before* you looked, the atom was *not* in one box or the other. It was in a superposition state simultaneously in both boxes. Therefore, assuming cats are not entities beyond the laws of physics, before you looked, the cat was in a superposition state equally alive and dead. It was not a sick cat. It was a perfectly healthy cat and a stone-dead cat at the same time.

Though the alive or dead condition of the cat did not exist as a physical reality until observed, the existence of the cat in the box *was* a reality. But, presumably, only because the existence of the cat in the box was observed by whoever put the cat there.

Since your looking collapsed the superposition state of the cat, are you guilty of killing the cat if you find it dead? Not really, assuming you didn't arrange the "hellish contraption" in the first place. You could not have chosen how the wavefunction of this entire system would collapse. The collapse into either the living or the dead state was random.

Here's something to ponder: Suppose the cat was placed in the box and the atom sent into the mirror system eight hours before you looked. The system evolves unobserved during those eight hours. If you find the cat alive, since it has gone eight hours without eating, you find a hungry cat. If you find a dead cat, an examination by a veterinary forensic pathologist would determine that the cat died eight hours ago. Your observation not only creates a current reality, it also creates the history appropriate to that reality.

You might consider all this absurd. Precisely Schrödinger's point! He concocted his cat story to argue that, taken to its logical conclusion, quantum theory was absurd. Therefore, he claimed, quantum theory must not be accepted as a description of what's *really* going on.

Notice that the enigma posed by Schrödinger's cat story is not quantum-theory-*neutral*, as was the enigma posed by our being able to freely choose to prove that an object had been either wholly in a single box or spread over two. Schrödinger's story displays an enigma posed by the quantum *theory*. Quantum theory describes the unobserved physical world as being a superposition of potentialities. This conflicts with our conscious observation telling us that the physical world is in a definite state.

The idea of a cat simultaneously alive and dead was, of course, as ridiculous to other physicists as it was to Schrödinger. But few worried about Schrödinger's demonstration of the theory's absurdity. The theory worked too well for mere absurdity to be a serious challenge.

No Fair Peeking

We soon come to the controversy that Schrödinger's story still raises. But first, if the cat is simultaneously alive and dead, can we somehow *see* it that way? No. Although we sketched a superposed live and dead cat, you'll never *see* a cat like that. Observation collapses the whole system to put the cat into either the living or the dead state. But what about just a peek? Can a tiny peek collapse the wavefunction of a whole cat?

Consider the tiniest possible peek. That might be bouncing a single photon off the cat through tiny holes in the box. With a single photon you can't learn much. But if that photon is blocked, telling us that the cat was standing, and therefore alive, that "look" would collapse the superposition state of the boxes to the cat in the alive state. Quantum theory tells us that *any* look, anything that provides information, collapses the previously existing state.

Suppose we see that the photon *does* come through the holes in the box. We then know that the cat is not standing. That "look" collapses the state of the boxes into a superposition of all states consistent with this observation. That superposition would include the state of a dead cat (and a fired Geiger counter), but also the state of an alive, but lying down, cat (and an unfired Geiger counter).

Wait a minute! Can't the *cat* observe whether or not the cyanide cork has been pulled, and therefore whether the atom entered its box? Don't cats qualify as observers and collapse wavefunctions? Well, if cats, what about mosquitoes? Viruses? Geiger counters? How far down do we go? We believe the two smart cats, the ones that live with each of us, are conscious observers. But how can we be sure?

Strictly speaking, all you know for sure is that *you* are a wavefunction-collapsing observer. The rest of us may be in superposition states governed by quantum mechanics and are collapsed to a specific reality only by your observation of us. Of course, since the rest of us look and act more or less like you, you trust that we also qualify as observers. (In chapter 15 we discuss the Many-Worlds interpretation of quantum mechanics, which suggests that we are *all* in superposition states.)

Although it can be seen as a logical extension of what quantum theory says, solopsistic talk of our being the only observer is just plain silly. As an alternative, some have seriously considered the possibility that quantum mechanics hints of a mysterious connection of *conscious* observation with the physical world. Eugene Wigner, one of the later developers of quantum theory and a winner of the Nobel Prize in Physics, created a version of the cat story suggesting an even stronger involvement of the conscious observer with the physical world than does Schrödinger's story.

Instead of a cat, Wigner contemplated having a friend stay unobserved in one of the boxes, a room. No cyanide this time. The firing Geiger counter just goes "click." His friend would mark an "X" on a pad if she hears a click. Wigner assumed that his friend's status as an observer was equal to his. He thus assumed that he did not collapse his friend's superposition state wavefunction when he opened the door and looked at her pad. She was never in a superposition state. He assumed that at least all humans have status as observers. Wigner *speculated* that collapse happens at the very last stage of the observation process, that his friend's human consciousness somehow collapsed the physical system's wavefunction. Going even further, he wondered whether human conscious awareness might actually "reach out"—in some unexplained way—and change the physical state of a system.

Such "reaching out" seems unreasonable to us. And eventually, Wigner thought so too. But you can't prove otherwise. All we know is that, some-place on the scale between big molecules and human awareness, there is this mysterious process of observation and collapse. Conceivably at least, it's indeed at the last step, at awareness. We explore some seriously pro-posed ideas regarding this in later chapters.

The Response to Schrödinger's Story

We've entered emotional territory. Most physicists squirm when their dis-cipline is associated with "soft" subjects such as consciousness. Some physicists claim that the cat story is nonsense, that it's misleading to even discuss such things. When reasonable people disagree on *testable* issues, they implicitly have the attitude, "I could be wrong." When refutation seems impossible, people are often *sure* of themselves. The practical impos-sibility of demonstrating, or refuting, Schrödinger's cat story is our case in point. Some physicists are even infuriated when the story is told. Stephen Hawking claims to "reach for my gun."

We'll give a more or less standard response to Schrödinger's story. First, though, a "truth in advertising" statement: Our sympathies are with Schrödinger's concern. Were that not so, we'd not be writing this book. Nevertheless, we'll present as strong an argument as we can that Schrödinger's cat story is irrelevant and misleading. For the next several paragraphs we try to take that point of view.

> Schrödinger's argument fails because it rests on the assumption that macroscopic objects can remain unobserved in a superposition state. For all practical purposes, any macroscopic object is **constantly** "observed." A big thing can't be isolated; it's always in contact with the rest of the world. And that contact *is* observation!
>
> It's ridiculous to even *imagine* that a cat could be isolated. Every macroscopic object anywhere near the cat effectively observes the cat. Photons are emitted by the warm cat to the walls of the box, and that means the box observes the cat. Take an extreme example: the moon! The moon's gravity, which pulls on the oceans to raise the tides, also pulls on the cat. That pull would be slightly different for a standing, alive cat than for a lying down, dead cat. Since the cat pulls back on the moon, the path of the moon is slightly altered depending on the posi-tion of the cat. Small as that effect might be, it is easy to calculate that in a tiny fraction of a millionth of a second the cat's wavefunction would

be completely entangled with the moon's, and thus with the tides and thus with the rest of the world. This entanglement *is* an observation. It collapses the superposition state of the cat in essentially no time at all.

Even looking back at the earliest stage of Schrödinger's story, you can see how absolutely meaningless it is. When an atom is sent into Schrödinger's boxes, its wavefunction becomes entangled with the enormously complex wavefunction of the macroscopic Geiger counter. The atom is therefore "observed" by the Geiger counter. Since something as big as a Geiger counter can't be isolated from the rest of the world, the rest of the world observes the Geiger counter and thus the atom. Entanglement with the world *constitutes* observation, and the atom collapses into one box or the other as soon as its wavefunction enters the box pair and encounters the Geiger counter. After that, the cat is either dead or alive. Period!

Even if you (needlessly!) bring consciousness into the argument, big things are constantly being observed if only because big things are constantly in contact with conscious beings.

If such arguments don't convince you that there's nothing to the cat story, here's a final put-down of Schrödinger's claim to having demonstrated a problem with quantum theory: *Do* the experiment! You'll always get the result quantum theory predicts; you'll always see either an alive or a dead cat.

Moreover, the Copenhagen interpretation makes it clear that the role of science is just to predict the results of observations, not to discuss some "ultimate reality." Predictions of what will happen are all we ever need. In this case, you'll find the cat alive half the time and dead half the time. Consciousness is irrelevant. The cat story raises a misleading non-issue.

We now no longer speak as a responder to Schrödinger's argument and return to our own voice. The physical impossibility of isolating an object as large as a cat to demonstrate it being in a superposition state is certainly correct. Schrödinger was, of course, fully aware of that difficulty. He would argue that such practical problems are beside the point. Since quantum theory admits no boundary between the small and the large, in principle *any* object can be in a superposition state. He (along with Einstein) rejected as defeatist the Copenhagen claim that the role of science is merely to predict the results of experiments, rather than to explore what's *really* going on.

No matter which side of this argument you favor, there are experts who'd agree with you.

Schrödinger's Cat Today

Seven decades after Schrödinger told his story, conferences almost every year address the quantum enigma and usually include discussions of consciousness. Reference to the cat story in professional physics journals increases. Two examples: One article, "'Schrödinger Cat' Superposition State of an Atom," demonstrated such a state for a microscopic system. In another, "Atomic Mouse Probes the Lifetime of a Quantum Cat," the "mouse" here is an atom and the "cat" is an electromagnetic field in a macroscopic resonant cavity. Though these were both serious and expensive physics projects, the titles illustrate how the physics discipline is inclined to approach the weirdness of quantum mechanics with a bit of humor.

Figure 11.2 Drawing by Aaron Drake, 2000.
© American Institute of Physics

Speaking of humor, here is a cartoon from the May 2000 issue of *Physics Today*, the most widely distributed journal of the American Institute of Physics. It would not likely have been published twenty years earlier.

Though the mysterious aspects of quantum mechanics are still hardly discussed in physics courses, the interest increases. A best-selling quantum mechanics text has a picture of a live cat on the front cover and a dead cat on the back—though there is very little talk of the cat inside. (Probably, the publishers, not the author, chose the cover design. But instructors choose the text, and we believe the allusion to the mystery appeals to younger faculty.)

Experimental studies of the mysterious aspects of quantum mechanics that would not have been proposed years ago, and would not have been funded if proposed, now get considerable attention. Increasingly large objects are being put into superposition states, put into two places at the same time. Austrian physicist Anton Zeilinger has done this with large molecules containing seventy carbon atoms, football-shaped "buckyballs." He's now setting up to do the same thing with mid-sized proteins and a virus. At a recent conference he was asked: "What's the limit?" His answer: "Only budget."

Truly macroscopic superpositions containing many billions of electrons have been demonstrated where each electron is simultaneously moving in two directions. Bose-Einstein condensates have been created in which each of several thousand atoms is spread over several millimeters. A 2003 American Institute of Physics news bulletin bore the headline "3600 Atoms in Two Places at Once." And here's the first sentence of a 2007 article in *Physical Review Letters*, a major physics research journal: "The race to observe quantum mechanical behavior in human-made nano-electromechanical systems (NEMS) is bringing us closer than ever to testing the basic principles of quantum mechanics." It gets harder to dismiss Schrödinger's concern by saying the weirdness only exists with the small things we can never actually see.

Perhaps hardest to accept is the claim that your observation not only creates a present reality but also creates a past appropriate to that reality—that, when your looking collapsed the cat to being either alive or dead, you also created the history appropriate to an eight-hour-hungry cat or to an eight-hour-dead cat.

The "delayed-choice experiment" suggested by quantum cosmologist John Wheeler, and discussed in chapter 7, comes closest to testing the backward-in-time aspect of quantum theory. It confirmed the prediction of quantum theory that an observation creates the relevant history.

Too bad Schrödinger isn't around to see the increasing interest in his cat. He felt that Nature was trying to tell us something, and that physicists should look beyond a pragmatic acceptance of quantum theory. He'd agree with John Wheeler: "Somewhere something incredible is waiting to happen."

12

Seeking a Real World
EPR

I think that a particle must have a separate reality independent of the measurements. That is, an electron has spin, location and so forth even when it is not being measured. I like to think the moon is there even if I am not looking at it.

—Albert Einstein

Schrödinger told his cat story by strictly applying quantum theory to the large as well as the small. His point was to ridicule the theory's claim that our observation *created* the reality we experience. That claim does seem crazy. Indeed, if someone on trial convinced the jury that he believed that his looking actually *created* the physical world, the jury would likely accept a plea of insanity.

The Copenhagen interpretation is, of course, more subtle. It does not deny a physically real world. It merely claims that objects of the *microscopic* realm lack reality before they are observed. Moons, chairs, and cats are real, if for no other reason than that macroscopic objects cannot be isolated, and are thus constantly observed. And that, according to Copenhagen, should be good enough. That was not good enough for Einstein.

At the 1927 Solvay conference, Einstein, by then the world's most respected scientist, turned thumbs down on the newly minted Copenhagen interpretation. He insisted that even little things have independent reality, whether or not anyone is looking. If quantum theory said otherwise, it had to be wrong. Niels Bohr, the Copenhagen interpretation's principal architect, rose to its defense. For the rest of their lives Bohr and Einstein debated as friendly adversaries.

Evading Heisenberg

Quantum theory has an atom being *either* an extended wave or a concentrated particle. If, on the one hand, you look and see it come out of a single box (or through a single slit), you show it to be a compact thing that had been wholly in a single box. On the other hand, you could have freely chosen to have the atom participate in an interference pattern, showing that it had been a spread-out thing *not* wholly in a single box. You can display either of two *contradictory* situations. The seemingly inconsistent theory is protected from refutation by the Heisenberg uncertainty principle. In this case, any look to see which box an atom comes out of would kick it hard enough to blur any interference pattern. You thus could not demonstrate a *logical* inconsistency.

To argue that quantum theory was inconsistent, and therefore wrong, Einstein would attempt to show that even though an atom participated in an interference pattern, it *actually* came through a single slit. To demonstrate this he would have to *evade* the uncertainty principle. (Ironically, Heisenberg attributed his original idea for the uncertainty principle to a conversation with Einstein.) Here's Einstein's challenge to Bohr at the 1927 Solvay conference:

Send atoms toward a two-slit diaphragm one at a time. Let the diaphragm be movable, say, on a light spring. Consider the simplest case, that of an atom that landed in the central maximum of the interference pattern (point A in figure 12.1). If that atom happened to come through the bottom slit, it had to be deflected upward by the barrier in order to land in the central maximum. In reaction, the atom would kick the diaphragm downward. And vice versa if the atom had come through the top slit.

By measuring the movement of the diaphragm after each atom had passed, one could know through which slit it went. This measurement could be made even after the atom was recorded as part of an interference pattern on a photographic film. Since one could thus know through which *single* slit each atom came, quantum theory was wrong in explaining the

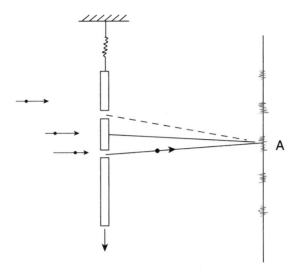

Figure 12.1 Atoms fired one at a time through a
movable two-slit barrier

interference pattern by claiming each atom to be a wave passing through
both slits.

Bohr readily pointed out the flaw in Einstein's reasoning: For Einstein's
demonstration to work, one would have to simultaneously know both the
diaphragm's initial position and any motion it might have had. The uncertainty principle limits the accuracy with which both the position and the
motion of an object can be simultaneously known. With simple algebra,
Bohr was able to show that this uncertainty for the slit diaphragm would
be large enough to foil Einstein's demonstration.

Three years later at another conference, Einstein proposed an ingenious
thought experiment claiming to violate an alternate version of the uncertainty principle. He would determine both the time at which a photon
exited a box and its energy, each with arbitrarily great accuracy. Einstein
would have a photon bounce back and forth in a box. A clock, opening a
door allowing the photon to leave, would record the precise time the photon
left. By leisurely weighing the box before and after the photon left, and
using $E = mc^2$, one could know the exact change in the energy of the system,
and thus the energy of the photon. Determining both the energy and the
time with arbitrary accuracy would then violate the uncertainty principle.

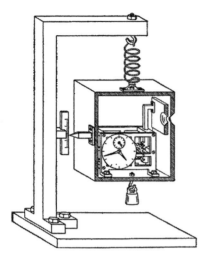

Figure 12.2 Bohr's drawing of Einstein's clock-in-the-box thought experiment. Courtesy HarperCollins

Figure 12.3 Albert Einstein and Niels Bohr at the 1930 Solvay Conference in Brussels. Photo by Paul Ehrenfest. Niels Bohr Archives

This one stumped Bohr through a sleepless night. But in the morning he embarrassed Einstein by showing that he had ignored his own theory of general relativity. To weigh the box, one must allow it to move in the Earth's gravitational field. According to general relativity, this would change the clock's reading by just enough to preclude a violation of the uncertainty principle. Years later, Bohr revisited his triumph with a nuts-and-bolts caricature of Einstein's photon-in-a-box experiment to illustrate that in any quantum experiment one must consider the macroscopic apparatus actually used.

The logic of Bohr's refutations of Einstein's thought experiments has been questioned. In chapter 10 we quoted Bohr as saying that "measuring instruments [must be] rigid bodies sufficiently heavy to allow a completely classical account of their relative positions and velocities." Was Bohr's application of quantum-mechanical uncertainty to the *macroscopic* slit diaphragm and his photon-box apparatus consistent with his requiring a "completely classical account" of the macroscopic measuring instruments?

At least Bohr seems to agree that quantum theory applies in principle to the large as well as the small. Only for all *practical* purposes do large things behave classically. Nevertheless, Bohr's arguments convinced Einstein that the theory was at least consistent, and that its predictions would always be correct. A humbled Einstein went home from the conference to concentrate on general relativity, his theory of gravity. Or so Bohr thought.

A Bolt from the Blue

Bohr was wrong. Einstein had not abandoned his attempt to fault quantum theory. Four years later (in 1935), a paper by Einstein and two young colleagues, Boris Podolsky and Nathan Rosen, arrived in Copenhagen. An associate of Bohr tells that "this onslaught came down upon us like a bolt from the blue. Its effect on Bohr was remarkable . . . as soon as Bohr heard my report of Einstein's argument, everything else was abandoned."

The paper, now famous as "EPR" for "Einstein, Podolsky, and Rosen," did not claim that quantum theory was wrong, just that it was *incomplete*. EPR argued that quantum theory did not describe the physically real world. It required an observer-created reality, only because it was not the whole story.

EPR showed that you could, in fact, know a property of an object without that property *ever* being observed. That property, EPR argued, was therefore *not* observer-created. Not being observer-created, it was a physical reality. If the quantum theory did not include such physical realities, it was an incomplete theory. Here's a classical analogy. It's the one that stimulated Einstein's EPR argument:

Consider two identical railroad cars latched together but pushed apart by a strong spring. Suddenly unlatched, they take off at the same speed in opposite directions. Alice, on the left in figure 12.4, is a bit closer to the cars' starting point than is Bob, on the right. Observing the position of the car passing her, Alice immediately knows the position of Bob's car. Having had no effect on Bob's car, Alice did not create its position. Not yet having observed his car, Bob did not create its position. The position of Bob's car was not observer created. It was therefore a physical reality. (A decade or so ago, physicists would talk of "observer A" and "observer B" in an EPR-type experiment. Today's friendlier convention is "Alice" and "Bob.")

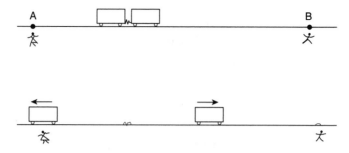

Figure 12.4 A classical analogy to the EPR argument

The conclusion arrived at in this Alice-and-Bob story is so obvious that it seems trivial. But replace the railroad cars with two atoms flying apart. Quantum theory describes them as spread-out wave packets. Their existence at *particular* positions is not a reality until the position of one of them is observed.

Unfortunately, there is a problem converting the easily visualized railroad car analogy to the quantum situation: The uncertainty principle forbids knowing both the initial speed and position of the cars accurately enough. We therefore skip EPR's ingenious but hard-to-visualize mathematical trick and go to the polarized photon version of EPR invented by David Bohm. Polarized photons are also worth exploring because the mysterious quantum influences eventually revealed by EPR-type experiments are most simply seen with photons. The actual demonstration of those quantum influences is the subject of our next chapter. But first, we need to see why Einstein considered them "spooky."

In the next few pages, we go over some of the physics of polarized light, and polarized photons, so that we can later present the profound EPR argument compactly. Even if you just lightly read these pages all the way down to the section headed "EPR," you can still appreciate the essence of Einstein's argument.

Polarized Light

Light, recall, is a wave of electric (and magnetic) field. Light's electric field can point in any direction perpendicular to the light's travel. In the top sketch of Figure 12.5, the light is going into the page with its electric field in the vertical direction. Such light is "vertically polarized." The lower sketch shows a horizontally polarized light wave. The direction of light's electric field is its polarization direction. From now on we'll just say "polarization" instead of "polarization direction."

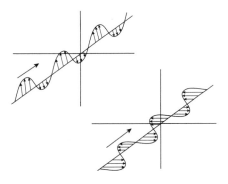

Figure 12.5 Vertically and horizontally polarized light

There is, of course, nothing special about the vertical and horizontal directions—other than that they are perpendicular to each other. It's just conventional to speak of "vertical" and "horizontal."

The polarization of light from the sun or a light bulb—most light, in fact—varies randomly. Such light is "unpolarized." Certain materials allow only light polarized along a particular direction to pass through. Such "polarizers" in sunglasses cut down glare by not transmitting the largely horizontally polarized light reflected from horizontal surfaces such as roads or water. But we will describe a different sort of polarizer.

The polarizers used in the most accurate experiments were transparent cubes formed of two prisms. We will refer to these cubes as "polarizers." These polarizers send light of different polarizations on two different paths. Light polarized parallel to a certain direction, the "polarizer axis," is sent on Path 1, and light polarized perpendicular to the polarizer axis is sent on Path 2.

Light polarized at an angle other than parallel or perpendicular to the polarizer axis can be thought of as composed of parallel and perpendicular polarized components. (It's the way a trip northeast can be thought of as composed of a trip with one component north and another east.) The parallel component of the light

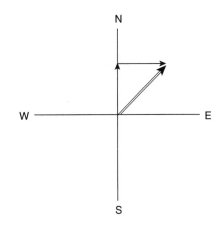

Figure 12.6 Traveling northeast as the sum of traveling north and then traveling east

goes on Path 1, and the perpendicular on Path 2. The closer the polarization is to parallel, the more light goes on Path 1.

Polarized Photons

Light is a stream of photons. Photon detectors can count individual photons. They can count millions per second. Our eyes, incidentally, can detect light as dim as a few photons per second.

Light polarized parallel to the polarizer axis is a stream of parallel-polarized photons. Each of them goes on Path 1 to be recorded by the photon detector on Path 1. Similarly, the Path 2 detector will record every photon with polarization perpendicular to the polarizer axis. The photons of ordinary, unpolarized light have random polarization. On encountering the polarizer, each is recorded by either the Path 1 or the Path 2 detector. In figure 12.7 we show a photon as a dot, its polarization as a double-headed arrow, the polarizer as a box, and the detectors as D1 and D2.

What about photons polarized at an angle other than parallel or perpendicular to the polarizer axis? Such photons have a *probability* of being recorded by either the Path 1 or the Path 2 detector. A photon polarized at

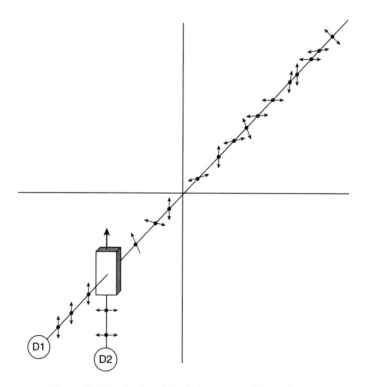

Figure 12.7 Randomly polarized photons sorted by a polarizer

forty-five degrees to the polarizer axis, for example, has equal probability of being recorded by either detector. The closer the polarization is to parallel to the polarizer axis, the greater its probability of being recorded by the Path 1 detector.

We were careful *not* to say that a photon at some angle other than parallel or perpendicular actually *went* on either path. It actually goes into a superposition state simultaneously having *both* polarizations and traveling simultaneously on *both* paths. A photon polarized at forty-five degrees, for example, goes equally on both paths.

But we never see partial photons. A detector clicks and records a whole photon, or it remains silent, indicating that no photon came to it. The situation of a photon coming on both paths is analogous to our atom being simultaneously in both boxes.

We could demonstrate that a photon had been in a superposition state on both paths by interference. Instead of a detector on each path, mirrors can direct each path through a second polarizer that recombines the parallel and perpendicular components of the photon to reproduce the original photon. Alter the length of *either* path, and the polarization of the emerging photon changes. This demonstrates that each photon was on *both* paths, in a superposition state having both polarizations, before being observed by a detector.

In saying photon detectors record photons, we're taking a Copenhagen interpretation stance. We're regarding the macroscopic photon detectors as observers. When a detector records the presence of a photon on a particular path, the superposition state collapses. What remains is the detector's recorded observation of the photon.

Einstein of course accepted the experimental results, but he did *not* accept this superposition-state business, where a photon had no particular polarization until it was observed. EPR would argue that each photon's polarization had to exist as a physical reality independent of its observation. Before we come to the EPR argument making this point, we must tell of photons in the "twin state."

Twin-State Photons

Atoms can be raised to excited states from which they return to their ground state by two quantum jumps in rapid succession, releasing

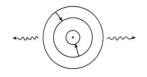

Figure 12.8 A two-photon cascade

two photons. Since there's nothing special about a particular direction in space, the photon polarization will be random.

But here's the *crucial* point: For certain special atomic states, the two photons flying off in opposite directions will always display the *same* polarization *as each other.* The photons are in a "twin state." If, for example, the photon going to the left is observed to have vertical polarization, its twin, the photon going to the right, will also be vertical.

The *reason* twin-state photons always exhibit the same polarization as each other doesn't matter here. (It's required to conserve angular momentum, and for the emission of twin-state photons, the initial and final atomic states have the same angular momentum.) The only important thing is that it is *demonstrably* true that their polarizations are always identical.

To demonstrate this, back to Alice and Bob, with photons instead of railroad cars. It's something actually done. A twin-state photon source is between Alice on the left and Bob on the right, as in figure 12.9. They each observe the polarization of twin-state photons with the axes of their polarizers oriented at the same angle, here both vertical. Their Path 1 and Path 2 photon detectors *randomly* click to record the simultaneous arrival of twin-state photons randomly polarized parallel or perpendicular to their polarizer axis. However, whenever Alice observes her Path 1 detector to record a photon, Bob *always* finds its twin on his Path 1. Whenever Alice observes her Path 2 detector to record a photon, Bob finds its twin on his Path 2.

We put no arrows on the photons in figure 12.9 because twin-state photons have no particular polarization, just the *same* polarization. We *did* put arrows on the photons in figure 12.7 because we considered the atoms emitting them as part of the macroscopic filament of the light bulb. Those atoms, and thus the polarization of the photons they emitted, were "observed" by that macroscopic object. Our twin-state photons were emitted by isolated atoms in a gas, thus not in contact with anything macroscopic.

Since the photons are twins, it might not *seem* strange that they always exhibit the same polarization. But it is strange. Let's play with an analogy: It's not surprising that identical-twin boys exhibit the same eye color. Identical twins are *created* with the same eye color. Consider, however, another property of the twins: the color of socks they choose to wear each day. Suppose, though far apart, whenever one twin chooses green, the

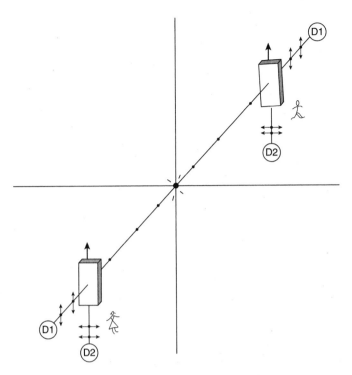

Figure 12.9 Alice and Bob with twin-state photons

other also chooses green that day, even though neither twin had information about his brother's sock-color choice. That would be strange, because the twins were not created with the same daily sock-color choice. Let's return to our twin-state photons.

Suppose that Alice, though far from Bob, is slightly closer to the photon source than is Bob. Her photon would be detected first. Whether it goes on her Path 1 or her Path 2, is completely random. However, if Alice's photon is recorded by her Path 1 detector, its twin will *always* be recorded by Bob's Path 1 detector.

Since Alice and Bob's photons moved away from the source in opposite directions at the speed of light, they separated at twice the speed of light. No physical force could ever connect the twin photons. The earlier detection of Alice's photon's random polarization could not *physically* affect Bob's photon. How, then, did Bob's photon instantly acquire the random polarization of Alice's?

It is not the *fact* that twin-state photons exhibit identical polarization that's weird. One might think that they were created not just with the *same* polarization, but also with a *particular* polarization. After all, our twin boys were not only created with the *same* eye color, but with a *particular* eye color, blue.

The weird thing is quantum theory's *explanation* of twin-state photons exhibiting identical polarization. According to quantum theory, *no* property is physically real until it is observed. Since the isolated emitting atom was left unchanged, it did not record, or "observe," the particular polarization of the twin-state photons it emitted. A particular polarization therefore did not exist as a physical reality. Thus, before Alice observed her photon's polarization, Bob's photon did not *have* a polarization. But *instantaneously* after Alice's distant observation, Bob's photon acquired a polarization—without any physical force involved. Weird.

Though Einstein's easily understood remark, "God does not play dice," is most quoted, quantum theory's denial of physical reality is what truly bothered Einstein. His less easily understood quip in the epigraph of this chapter, "I like to think the moon is there even if I am not looking at it," captures his *serious* objection. Though Einstein argued for an observer-independent real world, he was open-minded to a revolution. He wrote:

> It is basic for physics that one assumes a real world existing independently from any act of perception—but this we do not *know*. (Italics in original.)

EPR

The EPR paper that arrived in Copenhagen as a "bolt from the blue" was titled "Can Quantum-Mechanical Description of Physical Reality Be Considered Complete?" (Historians have attributed the missing "the" to the paper being worded by Podolsky, whose native Polish does not include articles.) The EPR paper presented a complex combination of the position and momentum of two particles. But we discuss it in the simpler, and more modern, way by talking in terms of photons.

Quantum theory has twin-state photons having *identical* polarizations, but does not include their *particular* polarization as a physically real property.

Nevertheless, quantum theory was claimed to be a *complete* theory, a theory describing *all* physically real properties.

To dispute this claim of completeness, EPR had to say what constituted a "physically real" property. Defining reality has long been, and still is, a debated philosophical issue. EPR offered a minimum condition for something to be a physical reality. EPR then argued that if such a physical reality existed, and was not described by quantum theory, the theory was incomplete. Here's the EPR definition:

> **If without in any way disturbing a system, we can predict with certainty . . . the value of a physical quantity, then there exists an element of physical reality corresponding to this physical quantity.**

Let's say the same thing in other words: If a physical property of an object can be known *without* its being observed, then that property could not have been created by observation. If such a property were not *created* by its observation, it must have existed as a physical reality *before* its observation.

Quantum theory does not contain *any* physical properties that are real in this sense. Therefore EPR needed to display only one such property as being physically real *before* it was observed to claim that quantum theory was incomplete.

That property would be the particular polarization of *one* of a pair of twin-state photons. EPR would argue that this particular polarization existed as a reality *before* its observation. Let's restate that argument concisely, even though it's essentially been made in our discussion of twin-state photons.

Back to Alice and Bob, with Alice a bit closer to the twin-state photon source than is Bob. She therefore receives her photon before Bob receives its twin. Suppose she observes a photon to be polarized vertically; it goes on her Path 1. She *immediately* knows that its twin, still on its way to Bob, has vertical polarization. She knows it will go on Bob's Path 1 when it reaches his polarizer.

In fact, it would be possible for Bob to trap his photon in a pair of boxes, one fed by his Path 1 and the other by his Path 2. After his photon is trapped, Alice could telephone Bob and tell him with certainty in which box he would find his photon.

Alice's observation of her photon could not have done anything *physical* to Bob's photon. Bob's photon moved toward Bob from the photon source at the speed of light. Since nothing can travel faster than light speed, nothing Alice could send at Bob's photon could ever catch up with it. She could not observe it. When Alice observed her photon, Bob's photon hadn't even gotten to him. Bob therefore could not have observed it.

Neither Alice nor Bob, nor anybody, observed the polarization of Bob's photon. Yet its unobserved polarization was known by Alice with certainty.

That does it! Alice knows with certainty the polarization of Bob's photon without observing it, without anyone observing it. This knowledge meets EPR's criterion for the polarization of Bob's photon being a physical reality. Since quantum theory does not include that physical reality, EPR claimed the theory was incomplete. The EPR paper concluded with the belief that a complete theory is possible. Such a complete theory would presumably give a *reasonable* picture of the world, a world existing independently of its observation.

Bohr's Response to EPR

When he received the EPR paper, almost a decade after the Copenhagen interpretation was developed, Bohr had not yet realized the implications of quantum theory to which EPR objected. He did not realize that the theory claimed that observation, in and of itself, *without any physical disturbance*, can instantaneously affect a remote physical system.

Bohr recognized Einstein's "bolt from the blue" as a serious challenge. He worked furiously for weeks to develop a response. A few months later he published a paper with exactly the same title as EPR: "Can Quantum Mechanical Description of Physical Reality Be Considered Complete?" (He even left out the "the.") While EPR's answer to the paper's title question was "no," Bohr's was a firm "yes." It was a largely philosophical response to EPR's scientific concern. Bohr countered EPR with what he called a "radical revision of our attitude as regards physical reality."

Here's an extract from Bohr's long response to EPR. It carries the essence of his complex argument:

> [The] criterion of physical reality proposed by Einstein, Podolsky and Rosen contains an ambiguity as regards to the meaning of the expression "without in any way disturbing a system." Of course there is in a case like that just considered no question of a mechanical disturbance of the system under investigation during the last critical stage of the measurement procedure. But even at this stage there is essentially the question of an *influence on the very conditions which define the possible types of predictions regarding the future behavior of the system.* Since these conditions constitute an inherent element of the description of any phenomenon to which the term "physical reality" can be properly attached, we see that the argument of the mentioned authors does not justify their conclusion that the quantum-mechanical description is essentially incomplete. (Italics in original.)

In his refutation of EPR, Bohr did not fault the logic of the EPR argument. He rejected their starting point, their condition for something being a physical reality.

EPR's reality condition tacitly assumes that if two objects exert no physical force on each other, what happens to one cannot in any way "disturb" the other. Let's be specific regarding Alice and Bob's twin-state photons: Alice, by observing her photon, cannot exert a physical force on Bob's photon, which is moving away from her at the speed of light. Therefore, according to EPR, she cannot have any effect on it.

Bohr agreed that there could be no "mechanical" disturbance of Bob's photon by Alice's observation. (*All* physical forces are included in Bohr's term "mechanical.") He nevertheless maintained that even without a physical disturbance, Alice's remote observation instantaneously "influences" what happens to Bob's photon. According to Bohr, this constitutes a disturbance violating the EPR condition for reality. Only after Alice *observed* her photon to be, say, polarized vertically was Bob's photon polarized vertically.

How did Alice's observation affect Bob's photon? Can what is done at a distant place, even on a faraway galaxy, instantaneously *cause* something to happen here? Strictly speaking, we should not say her observation "affected" Bob's photon or "caused" its behavior because no physical force was involved. We use the mysterious term sanctified by Bohr: Alice "influenced" Bob's photon's behavior.

Though Alice instantaneously influenced Bob's photon, she cannot communicate any information to Bob faster than the speed of light would allow. Bob always sees a series of random photon polarizations. Only when Alice and Bob come together and compare their results do they see that whenever she saw a photon polarized vertically, so did he; whenever she saw a photon polarized horizontally, so did he.

To defend quantum theory despite this "nonphysical" influence, Bohr later redefined the goal of science. That new goal is not to *explain* Nature, but only to describe what we can *say* about Nature. In his early debates with Einstein, Bohr defended quantum theory by arguing that any observation *physically* disturbs what you examine by enough to prevent any refutation of quantum theory. This has been called a "doctrine of physical disturbance." Since Alice's observation supposedly only changes what can be *said* about Bob's photon, Bohr's response to EPR has been called a "doctrine of semantic disturbance."

Is all this confusing? You bet! There is no way that EPR, and Bohr's response to it, can be correctly stated that does not either confuse or sound mystical.

Einstein rejected Bohr's response. He insisted that there was a real world out there. The goal of science must be to *explain* Nature, not just tell what we can *say* about Nature. A photon displayed a particular polarization, Einstein claimed, because that photon actually *had* that polarization. He insisted that objects have physical properties independent of their observation. If any such properties, later called "hidden variables," were not included in quantum theory, the theory was incomplete. Einstein derided Bohr's remote "influences" as being "voodoo forces" and "spooky actions." He could not accept such things as part of the way the world works, saying: "The Lord God is subtle, but malicious He is not."

We should be clear that Bohr and Einstein would agree on the actual *results* of an EPR experiment, the Alice-and-Bob observations we described. They would just *interpret* those results differently. That's why no one actually *did* an EPR experiment. All physicists knew what the result would be. The Einstein-Bohr argument was regarded as "merely philosophical."

Einstein was forever dubious about quantum theory; Bohr was its staunchest defender. It's fair to speculate why Einstein and Bohr held so strongly to their philosophical positions. Recall that for almost twenty years the physics community rejected young Einstein's quantum proposal that light came as photons. It was called "reckless." By contrast, young Bohr's proposal of a quantum effect brought him immediate acclaim. How much did their early professional experiences with quantum theory shape their lifelong attitudes towards it?

Einstein thought physicists would reject Bohr's argument refuting EPR. He was wrong. Quantum theory worked too well. It provided a basis for rapid advance in physics and its practical applications. Working physicists had little inclination to deal with philosophical issues.In the thirty years after the 1935 publication of EPR, it was essentially ignored. It was cited on average only once per year. Since Bell's theorem (treated in our next chapter), that's changed. Between 2002 and 2006, EPR was cited over 200 times a year, and interest increases. EPR is today probably the most cited physics paper from the first half of the twentieth century.

In the two decades he lived after EPR, Einstein never wavered in his conviction that there was more to say than quantum theory told. He urged his colleagues not to give up the search for the secrets of "the Old One." But he may have become discouraged. In a letter to a colleague, he wrote: "I have second thoughts. Maybe God *is* malicious."

Experiments motivated by EPR have now established the actual existence of the "spooky actions" that bothered Einstein. They are now referred to as "entanglements." Industrial laboratories work with entanglements as the basis of quantum computers. They are nevertheless still spooky. They are the subject of our next chapter.

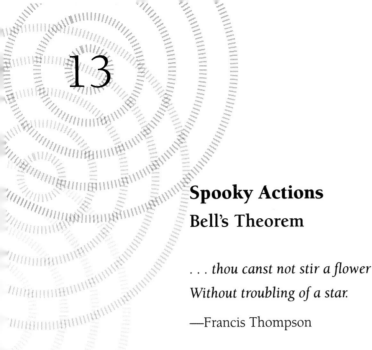

13

Spooky Actions
Bell's Theorem

. . . thou canst not stir a flower
Without troubling of a star.

—Francis Thompson

Physicists paid little attention to EPR, or to Bohr's response. Whether or not quantum mechanics was complete mattered little. It *worked*. It never made a wrong prediction, and practical results abounded. Who cared if atoms lacked "physical reality" before being observed? Working physicists had little time for an unanswerable "merely philosophical question."

Shortly after EPR, physicists, giving their attention to the Second World War, developed radar, the proximity fuse, and the atom bomb. Then came the politically and socially "straight" 1950s. In physics departments a conforming mind-set meant that an untenured faculty member might endanger a career by seriously questioning the orthodox interpretation of quantum mechanics. Even today, it's best to explore the meaning of quantum mechanics while also working a "day job" on a mainstream physics topic. However, since a theorem proven by John Bell, physicists, especially younger physicists, show increasing interest in what quantum mechanics is telling us.

Bell's theorem has been called "the most profound discovery in science in the last half of the twentieth century." It has rubbed physics' nose in the weirdness of quantum mechanics. Bell's theorem and the experiments it stimulated answered what was supposedly a "merely philosophical question" in the laboratory. We now know Einstein's "spooky actions" actually exist. Even events at the edge of the galaxy instantly influence what happens at the edge of your garden. We quickly emphasize that such influences are undetectable in any normally complex situation.

Nevertheless, What are now called "EPR-Bell influences," or entanglement, now get attention in industrial laboratories for their potential to allow incredibly powerful computers. They already provide the most secure encryption for confidential communication. Bell's theorem has renewed interest in the foundations of quantum mechanics, and dramatically displays physics' encounter with consciousness.

John Stewart Bell

John Bell was born in Belfast in 1928. Though no one in the family had ever had even a secondary school education, his mother promoted learning as the way to the good life, in which you "could wear Sunday suits all week." Her son became an enthusiastic student and, by his own assessment, "not necessarily the smartest but among the top three or four." Eager for knowledge, Bell spent time in the library instead of going off with the other boys, which he would have done had he been, he says, "more gregarious, more socially adequate."

Early on, philosophy attracted Bell. But finding each philosopher contradicted by another, he moved to physics, where "you could reasonably come to conclusions." Bell studied physics at Queen's, the local university. In quantum mechanics, the philosophical aspects interested him most. For him, the courses concentrated too strongly on the practical aspects of the theory.

Nevertheless, he finally went to work in an almost engineering role, the design of particle accelerators, eventually at CERN (the European Center for Nuclear Research) in Geneva. But he also produced important work in theoretical physics. He married a fellow physicist, Mary Ross. Though they worked independently, Bell writes that in looking through his collected papers, "I see her everywhere."

At CERN, Bell concentrated on the mainstream physics he felt he was paid to do, and of which his colleagues approved. He restrained his interest in the weirdness of quantum mechanics for years. Eventually an opportunity to explore these ideas came in 1964 on sabbatical leave. Bell tells that, "Being away from the people who knew me gave me more freedom,

Figure 13.1 John Bell. © Renate Bertlmann 1980.
Courtesy Springer Verlag

so I spent some time on these quantum questions." The momentous result we now call "Bell's theorem."

I (Bruce) shared a taxi and conversation with John Bell in 1989 on the way to a small conference in Erice, Sicily, which focused on his work. At the conference, with wit, and in his Irish voice, Bell firmly emphasized the depth of the unsolved quantum enigma. In big, bold letters on the blackboard he introduced his famous abbreviation, FAPP, "for all practical purposes," and warned against falling into the FAPPTRAP: accepting a merely FAPP solution for the enigma. As department chair at the time, I was able to invite Bell to spend some time in our physics department at the University of California, Santa Cruz, and he tentatively accepted. But the next year John Bell suddenly died.

Bell's Motivation

Recall that EPR accepted all the predictions of quantum theory as correct. They challenged its completeness. They claimed that the theory's observation-created reality arose from its not including the physically real properties of objects, "hidden variables." EPR's argument started with the "obvious," implicit assumption that only physical forces could affect the behavior of objects. Since no physical effect can travel faster than the speed of light, two objects could therefore be separated so that the behavior of one could not affect the other in a time less than it would take light to go from one to the other. EPR's argument for reality thus *assumed* separability.

In refuting EPR, Bohr *denied* separability. He claimed that what happened to one object could indeed "influence" the behavior of another *instantaneously*, even though no physical force connected them. Einstein derided Bohr's "influences" as "spooky actions," "*spukhafte Fernwirkung*," in his original German.

For thirty years, no experimental result could decide between Einstein's supposedly physically-real hidden variables and Bohr's instantaneous "influences." Moreover, physicists tacitly accepted a mathematical theorem claiming to prove that it was *impossible* for a theory that included hidden variables to reproduce the predictions of quantum theory. That theorem undercut Einstein's argument for hidden variables.

While Bell enjoyed sabbatical freedom to explore such things, he was struck by a *counterexample* to the no-hidden-variables theorem. He discovered that, twelve years earlier, David Bohm had developed a theory that *included* hidden variables yet also reproduced the predictions of quantum mechanics. "I saw the impossible done," said Bell.

After discovering where the no-hidden-variables theorem went wrong, Bell pondered: Since hidden variables *might* exist, do they *actually* exist? How would a world with such real, observation-independent properties differ from the world that quantum theory describes? Bell wanted to understand what the quantum calculations physicists do actually *mean*. He wrote: "You can ride a bicycle without knowing how it works. . . . In the same way we [ordinarily] do theoretical physics. I want to find the set of instructions to say what we are really doing."

Bell's Theorem

Because EPR's argument challenged none of quantum theory's *predictions*, EPR did not confront the theory with an experimental challenge. Bell did challenge the theory. He derived an experimentally *testable* prediction that *had to be* true in *any* world that included an observation-independent reality and separability. Quantum theory *denies* such reality and separability. Bell's testable prediction was a "straw man" he created for experiments to try to knock down. Should Bell's straw man *survive* experimental challenge, quantum theory would be shown wrong.

Bell's theorem in a nutshell: Suppose our world to have physically real properties that are *not* created by their observation. Further, suppose that objects can be separated from each other so that what happens to one cannot instantaneously affect the other. (For short, we call these two suppositions "reality" and "separability.") From *only* these two premises—both assumed in classical physics but denied by quantum theory—Bell deduced that certain observable quantities could not be larger than certain other observable quantities. This *experimentally testable* conclusion of Bell's theorem, which *must* be true in any world with reality and separability, is "Bell's inequality."

If Bell's inequality is shown to be false in *any* situation, one or both of the premises logically leading to it (reality and separability) *must* be false. Therefore, if in our actual world Bell's inequality is *ever* violated, *no* objects with reality and separability can exist in our actual world. (Bell expected the inequality to be violated, as quantum theory predicts.)

The most commonly observed quantities used in testing Bell's inequality are the rates at which twin-state photons display different polarizations when polarizers are set at different angles. But for the moment, let's be more general.

All this is pretty abstract. Philosophers and mystics have talked of reality and separability (or its opposite, "universal connectedness") for millennia. Quantum mechanics puts these issues squarely in front of us. Bell's theorem allows them to be tested.

In what we will call a "reasonable" world, objects have physically real properties (not merely properties created by their observation).

Moreover, in such a reasonable world, objects are separable. That is, they affect each other only by physical forces, which cannot travel faster than the speed of light (not by "spooky actions" traveling infinitely fast). The Newtonian world described by classical physics is, in this sense, a reasonable one. The world described by quantum physics is *not*. Bell's theorem allows a test to see whether perhaps it's just quantum theory's *description* of our world that's unreasonable, and that our actual world is in fact a reasonable one.

We won't go for suspense. When the experiments were done, Bell's inequality was violated. Assumptions of reality and separability yielded a *wrong* prediction in our actual world. Bell's straw man was knocked down, as Bell expected it would be. Our world therefore does *not* have both reality and separability. It's in this sense, an "unreasonable" world.

We immediately admit not understanding what the world lacking "reality" might mean. Even what "reality" itself might mean. In fact, whether or not reality is indeed required as a premise in Bell's theorem is in dispute. However, we need not deal with that right now. For *our* derivation of a Bell inequality, we assume a straightforward real world. Later, when we discuss the consequences of the violation of Bell's inequality in our actual world, we'll define a "reality" implicitly accepted by most physicists. It will leave us with a strangely connected world.

Derivation of a Bell Inequality

We offer a derivation of a Bell inequality with objects something *like* twin-state photons. We will call our objects "fotons." Each of our twin-state fotons has a physically real polarization angle, just called its "polarization." Moreover, the twin fotons can be separated so that what happens to one cannot instantaneously affect the other. Our fotons are clearly *not* the photons of quantum theory, which *denies* such reality and separability.

Do the photons that make Geiger counters click in our actual world have the quantum-theory-denied reality and separability of our fotons? That's something experiments with actual photons must decide.

To be concrete, we present a specific mechanical picture. However, the logic we use *in no way depends on any aspect of this mechanical model* except the reality of each foton's polarization and its separability from its twin.

Bell's mathematical treatment was completely general. It did not even specify photons.

If you only skim our pictorial derivation of a Bell inequality and just accept the result, you will not be *much* hampered in understanding the rest of the book. For a fast, first reading, you might even skim all the way down to "An intentionally ridiculous story" and Figure 13.6.

An Explicit Model

In figures 13.2, 3, 4, and 5, we present a specific mechanical picture. To display each foton's assumed polarization as graphically real, we show a foton as a stick, and the angle of the stick is its polarization. Picturing fotons as sticks necessarily displays properties beyond their polarization. These properties, a stick's length or width, for example, are irrelevant to our derivation. Only the foton's physically real polarization is relevant. This is our reality assumption. Its polarization determines which path a foton takes upon encountering a "polarizer."

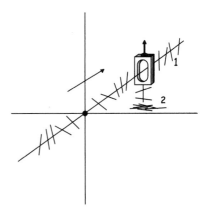

Figure 13.2 Model of stick photons and oval polarizer

A "polarizer" in this mechanical model is a plate with an oval opening whose long dimension is the "polarizer axis." A foton whose polarization is close to the polarizer axis will pass through the polarizer to go on Path 1. One whose polarization is not close will hit the polarizer to go on Path 2.

This mechanical model could in principle, but need not, account properly for all the behavior of actual polarized light. Our logic depends on nothing about these fotons except their reality and separability.

We will describe four Alice-and-Bob thought experiments. They are much like the EPR experiment described in chapter 12. (In fact, Bell's theorem experiments are sometimes loosely referred to as EPR experiments.) But there's a big difference: In the EPR case, Einstein's "hidden variables" and Bohr's "influences" led to the *same* predicted experimental outcome. The disagreement between Bohr and Einstein was only a difference of *interpretation*. In our model, and in the actual Bell's theorem experiments, the outcomes for Einstein's "hidden variables" and Bohr's "influences" are different.

In each of our four Alice-and-Bob experiments, twin-state fotons with identical, but random, polarizations are emitted in opposite directions from a source between Alice and Bob. Since twin-state fotons fly apart from each other at the speed of light, nothing physical can get from one experimenter to the other in the time between the fotons arriving at their respective polarizers. Therefore, what happens to one of our fotons at one polarizer cannot affect its twin at the other polarizer. This is our separability assumption.

As in the EPR case, Alice and Bob identify fotons as being twins by their simultaneous arrival times and keep track of whether their Path 1 or Path 2 detector recorded each foton.

Experiment I

In this first experiment, as in the original EPR experiment, Alice and Bob each have their polarizer axes aligned vertically. They record a "1" every time their Path 1 detector records a foton and a "2" every time their Path 2 detector records one. They each end up with a string of random 1s and 2s.

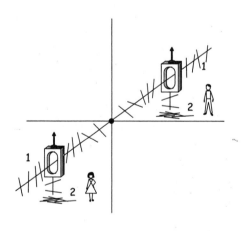

After recording a large number of fotons, Alice and Bob come together and compare their results. They find their data streams identical. Bob's foton took the same path at his polarizer as its twin did at Alice's. This confirms that simultaneously arriving fotons are twins.

Alice and Bob expected this perfect matching. A pair of twin-state fotons indeed *had* identical polarization. In this model with reality the fotons were *created* with identical polarizations. (In quantum theory, on

Figure 13.3 Experiment I: Polarizers are aligned, and Alice's and Bob's data are identical

the other hand, where polarization is *observer* created, the matching must be explained by an "influence" instantaneously exerted on a photon by the observation of its remote twin.)

Experiment II

This is the same as Experiment I, except this time Alice rotates her polarizer by a small angle we'll call Θ (the Greek letter theta). Bob keeps his polarizer axis vertical.

Both experimenters take the same kind of data once more. The polarization of fotons is unaffected by Alice's choice of a new polarizer axis. Therefore, some fotons that *would have* gone through Alice's polarizer on her Path 1, had she not rotated it, now go on her Path 2, and vice versa. By our separability assumption, Bob's fotons are unaffected by Alice's polarizer rotation, or by which path their twins took at Alice's polarizer.

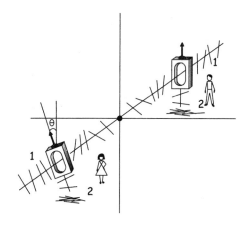

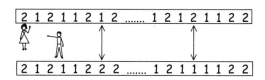

Figure 13.4 Experiment II: Alice's polarizer is rotated, causing mismatches

Alice and Bob, coming together this time to compare their random data streams, find some mismatches. Mismatches occur, for example, when some of Alice's fotons, which would have gone on her Path 1 had she not rotated her polarizer, went on her Path 2. But their twins at Bob's polarizer still went on his Path 1. The percentage of mismatches would be small for small Θ. Let's say that Alice changed what would have happened for 5% of her fotons. She thus caused a mismatch rate of 5%.

Experiment III

This is exactly the same as Experiment II, except that Bob rotates his polarizer by the angle Θ, while Alice returns hers to the vertical. Since the situations are symmetrical, the mismatch rate again would be 5%, assuming that the number of foton pairs was large enough that statistical error was negligible.

Experiment IV

This time both Alice and Bob each rotate their polarizers by the angle Θ. If they each rotated in the same direction, it would be the same as no rotation at all; their polarizers would still be aligned. So they each rotate their polarizers by Θ in opposite directions.

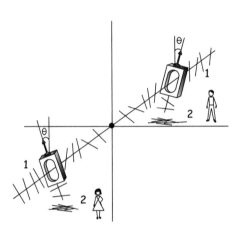

Alice, rotating her polarizer by Θ, changes the behavior of her fotons by the same amount as in Experiment II. She changes what would have happened to 5% of her fotons. The situation is symmetrical. Bob's polarizer rotation by Θ changes the behavior of 5% of his fotons from what would have happened.

Since Alice and Bob each changed the behavior of 5% of their fotons, and since every change could show up as a mismatch when their data streams are compared, we might expect a mismatch rate as high as 10%. There is no way to get a *greater* mismatch rate in a statistically large sample.

Figure 13.5 Experiment IV: Both Alice's and Bob's polarizers are rotated, and mismatches are due to both rotations

We might, however, get a *smaller* mismatch rate. Here's how: It's likely that for some pairs of twin-state fotons, both Alice and Bob *each* caused their twin to change its behavior. The two fotons of such twin-state pairs would thus behave identically. The data for such twin-state pairs would not be recorded as mismatches.

As an example of such a double change of behavior, consider almost vertical twin-state fotons that would have both gone on Path 1 at Alice's and Bob's polarizers had their polarizer axes both remained vertical. If

Alice and Bob each rotated their polarizers in opposite directions, as they did in Experiment IV, they could each send this pair of twins on their Path 2. They would thus not record this double change as a mismatch.

Because of such double changes, when Alice and Bob compare their data streams in Experiment IV, the mismatch rate will likely be *less* than the 5% error rate Alice alone would cause *plus* the 5% error rate that Bob alone would cause. In Experiment IV the mismatch rate they will see is likely *less* than 10%. In a statistically large sample it *cannot* be greater.

That's it! We've derived a Bell inequality:

> **The mismatch rate when both polarizers are rotated by Θ (in opposite directions) is equal to, or less than, twice the mismatch rate for the rotation by Θ of a single polarizer.**

Since space is the same in all directions, a rotation of the two polarizers in opposite directions by Θ is equivalent to a rotation of only one by 2Θ. Thus rotating a single polarizer in one experiment by Θ and in a second experiment, by 2Θ can demonstrate the same inequality. The Bell inequality would then state: A rotation by 2Θ cannot produce more mismatches than twice those for a rotation by Θ.

Here's an intentionally ridiculous story to emphasize that the *only* assumptions in our derivation of a Bell inequality were reality and separability. Instead of talking of stick-like fotons and oval polarizers, we could have said that each foton is a steered by a little "foton pilot" and that a polarizer is just a traffic sign indicating an "orientation" with an arrow. The foton pilot carries a travel document instructing him to steer his "foton" on Path 1 or Path 2 depending on the traffic sign. The hidden variable is now

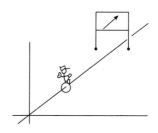

Figure 13.6 The photon pilot

the physically real instruction printed on the pilot's travel document. His sister, piloting the foton's twin, follows her identical instructions at the traffic sign she encounters with no regard for the behavior of her brother. This model yields the same Bell inequality. Only reality and separability need be assumed.

Suppose *actual* experimental data *violated* the Bell's inequality we just derived. That is, suppose in laboratory experiments with actual twin-state photons the mismatch rate for the rotation of both polarizers was *greater* than twice the mismatch rate for the rotation of a single polarizer. Since a Bell's inequality, saying it could *not* be greater, was deduced assuming *only* reality and separability, its violation would mean that one or both of those assumptions *had* to be wrong in the actual world. It would mean that our actual world lacked either reality, or separability, or both. We will see that a violation in any one case (actual twin-state photons, for example) means a lack of reality or separability for everything such photons could possibly interact with. In principle, that's anything. (We use the adjective "actual" rather than the tricky word, "real," to refer to the world we live in and the photons we deal with.)

Were Bell's inequality *not* violated, quantum theory, which *predicts* a violation, would have been shown wrong. But nothing would be proven about reality or separability. Incorrect assumptions can lead to *some* correct predictions. In fact, in some situations, Bell's inequality is not violated. Its violation in *any* situation is sufficient to deny that our actual world has both reality and separability.

The Experimental Tests

In 1965, when Bell's theorem was published, it was a mild heresy for a physicist to question quantum theory or even to doubt that the Copenhagen interpretation settled all philosophical issues. Nevertheless, as a physics graduate student at Columbia University in the late 1960s, John Clauser was intrigued.

Off to Berkeley as a postdoc to work on radio astronomy with Charles Townes, Clauser presented his idea for a test of Bell's inequality. Townes released him from his commitment to work on astronomy, and even continued his financial support. With borrowed equipment, Clauser and a graduate student measured what we have called the "mismatch rate" for twin-state photons with polarizers set at different angles with respect to each other. They, in essence, did the Alice-and-Bob experiments. They found Bell's inequality *violated*. Violated in just the way quantum theory predicts.

To avoid a common misstatement, we emphasize that Bell's *inequality* was violated. Bell's *theorem*, the derivation of the inequality from the assumptions of reality and separability, is a mathematical proof not subject to experimental test.

Exactly What Does Quantum Theory Predict?

The actual *amount* by which quantum theory predicts Bell's inequality to be violated requires a rather complicated calculation, one that is not particularly relevant for our discussion. However, for those who want to explore this point, we'll say a bit more, but the following paragraph *can well be skipped.*

A semi-classical calculation considering light as an electric field gives the correct answer for the mismatch rate, even though it cannot deal with the photon correlations needed to establish the *meaning* of Bell's inequality. We note here, without much explanation, the following facts: (1) Alice's observation of an actual photon going through her, say, vertical polarizer means its twin at Bob's polarizer will be vertical. (2) The fraction of light intensity (or photons) *not* going through Bob's polarizer, the mismatch rate, is proportional to the square of the component of electric field perpendicular to his polarizer axis. (3) This fraction is proportional to the square of the sine of the angle Θ of Bob's polarizer to Alice's (Malus's law). Thus, the actually observed mismatch rate, and that given by quantum theory, is proportional to $\sin^2(\Theta)$. (4) The Bell inequality we derived thus states: $2\sin^2(\Theta) \geq \sin^2(2\Theta)$. Try this for $\Theta = 22.5°$, $2\Theta = 45°$. We get $0.3 \geq 0.5$. Very wrong. We thus see that in the actual world, Bell's inequality can be strongly violated. We repeat: this paragraph can be skipped.

The Bottom Line for the Experimental Results

Clauser's experiments ruled out what is sometimes called "local reality," or "local hidden variables." The experiments showed that properties of our world either have only an observation-created reality *or* that there exists a connectedness beyond that mitigated by ordinary physical forces, or both.

Clauser writes: "My own . . . vain hopes of overthrowing quantum mechanics were shattered by the data." Instead, he confirmed quantum theory's predicted violation of Bell's inequality. In his experiments, quantum theory survived its most serious challenge in decades.

We can never be certain that any scientific theory is correct. Some day a better theory might supersede quantum theory. But we now know that any such better theory must also describe a world that does not have both reality and separability. Before Clauser's result, we could not know this.

Unfortunately for Clauser, in the early 1970s investigation of the foundations of quantum mechanics was not yet considered proper physics in most places. When he sought an academic position (including an opening in our own department at the University of California, Santa Cruz), his work was met with scorn. "What has he done besides checking quantum theory? We all *know* it's right!" was a typical misunderstanding of Clauser's accomplishment. Clauser got a job in physics, but not one in which he could participate in the wide-ranging investigations his work launched.

A decade later Alain Aspect, in France, upgraded Clauser's test of Bell's inequality. With the extremely fast electronics available, he could assure that the time between the detection of one photon and the detection of its twin would be less than the time it would take light to get from one detection apparatus to the other. Therefore, since no physical force can propagate faster than the speed of light, no physical effect of one photon's observation could possibly affect that of its twin. This closed a loophole in Clauser's experiment, whose electronics were not quite fast enough.

Aspect reports that when he told Bell of his plans, Bell's first question was, "Do you have tenure?" Exploration of the foundations of quantum mechanics was more acceptable than it was a decade earlier, but it was still to be undertaken with caution for one's career. Aspect's eventual results not only closed the loophole in Clauser's experiment but produced strikingly strong confirmation that Bell's inequality was violated in precisely the way quantum theory predicted. If John Bell had not died, Bell, Clauser, and Aspect might well share a Nobel Prize.

The Aspect result will not be the end of the story. In Bell's words:

> It is a very important experiment, and perhaps it marks the point where one should stop and think for a time, but I certainly hope it is not the end. I think the probing of what quantum

mechanics means must continue, and in fact will continue, whether we agree or not that it is worth while, because many people are sufficiently fascinated and perturbed by this that it will go on.

More than two decades after his prediction, today's probing of the meaning of quantum mechanics confirms Bell's insight.

Where Does a Violation of Bell's Inequality Leave Us?

Reality First

"Reality" has been our shorthand term for the existence of physically real properties *not* created by their observation. Quantum theory does not include such reality. The nature of physical reality has been argued about at least since Plato's day in 400 BC. And it still is. In particular, the question of whether or not Bell's theorem actually assumes reality is debated. (With reality *not* one of the theorem's premises, the experimental results would deny separability for our actual world, not just deny the coexistence of reality *and* separability.)

(This technical paragraph may be skipped. Bell's mathematical derivation includes a symbol, λ, the Greek letter lambda, representing *everything* existing in the past that could possibly affect outcomes at Alice's site without also affecting those at Bob's site, and vice versa. If λ includes actual polarizations for the incoming photons, the reality of the properties of specific objects would be a premise of Bell's theorem. If, however, λ refers to *all possible* aspects of the observations, including, say, aspects of the polarizers, but not photon polarizations as individual properties, the reality postulate as applied to specific objects is deniable. In our pictorial derivation of a Bell inequality, where Alice and Bob observed foton polarization angles, objects had a straightforward reality.)

To pursue a reality argument, let us now assume *complete* separability, that *nothing* Alice could do (*including* any faster-than-light "influences") could have *any* affect on the outcomes at Bob's polarizer. Let us also accept a definition of physical reality in the spirit of EPR: If a property of an object can be known without any observation of it, that property was not *created* by observation. It thus existed as a physical reality.

This definition of physical reality carries philosophical baggage. But so does any other. It is, however, a definition implicitly accepted by most physicists, (and probably by most people).

Now a bit of logic: 1) Assuming separability, and given the results of the *actual* world version of Alice and Bob Experiment I, an EPR-like reality is established. 2) The actual world versions of Alice and Bob Experiments II, III, and IV establish that separability *and* reality cannot *both* exist in the actual world. However, according to 1), if we have separability, we *must* have reality, but, according to 2), we can't have both. We thus *rule out* separability in our actual world.

Separability

"Separability" has been our shorthand term for the ability to separate objects so that what happens to one in *no way* affects what happens to others. Without separability, what happens at one place *can* instantaneously affect what happens far away–even though no physical force connects the objects. Bohr accepted this strange quantum theory prediction as an "influence." But to Einstein, an effect coming about without an actual physical force was a "spooky action."

That our actual world does not have separability is now generally accepted, though admitted to be a mystery. In principle, any objects that have ever interacted are forever entangled, and therefore what happens to one influences the other. Experiments have now demonstrated such influences extending over more than one hundred kilometers. Quantum theory has this connectedness extending over the entire universe. Designers of quantum computers have demonstrated these influences connecting the almost macroscopic logic elements for prototype quantum computers.

Quantum connectedness can, in principle at least, extend beyond the microscopic to the macroscopic. The lack of separability of any two objects, for example, twin-state photons, establishes a lack of separability generally. Consider, for example, Schrödinger's "hellish contraption" constructed so that a twin-state photon entering the cat box would trigger a cyanide release if that photon displayed vertical polarization but would not trigger the release if it displayed horizontal polarization. The fate of the cat would be then determined by the remote observation of the polarization of the photon's twin. Of course, since the polarization of the remote photon was random, so is the fate of the cat. There is no remote control.

We talk in terms of twin-state photons because that situation is readily described and subject to experimental test. We extend the idea to Schrödinger's cat because that situation is easy to describe, though essentially impossible to demonstrate. In principle, however, any two objects that have ever interacted are forever entangled. The behavior of one instantaneously influences the other, and the behavior of everything entangled with it. Since truly macroscopic objects are almost impossible to isolate, they rapidly become entangled with everything else in their environment. The effect of such complex entanglement generally becomes undetectable. Nevertheless, there is, in principle, a universal connectedness whose meaning we have yet to understand. We can indeed "see the world in a grain of sand."

Does infinitely fast quantum entanglement conflict with special relativity, which holds that no physical effect can travel faster than the speed of light? Special relativity is basic to much of physics, and every test of relativity accurately accords with the theory's predictions. Nevertheless, some of relativity's underlying assumptions may be challenged by the lack of separability. A recent *Scientific American* article, for example, is titled "A Quantum Threat to Special Relativity." In any event, no information, message, or causal effect can be sent from one observer to another faster than the speed of light. Bob just records a random series of 1s and 2s. Only when he compares his data with Alice's can they see the EPR-Bell correlations.

In discussing EPR (or Experiment I) we said Alice looked first and influenced Bob's photon. We've been asked, "What if Alice and Bob looked at the same time?" According to special relativity, for some observers moving with respect to Alice and Bob, it would be Bob who looked first, not Alice. Quantum theory just says that the two observers, with their polarizers aligned, will see the *same* polarization, and see correlations in their data at other angles.

Induction and Free Will

The quantum connectedness forces us to examine issues that once seemed beyond the realm of physics. Bell's theorem, more explicitly than anything in classical physics, rests on the validity of inductive reasoning, and even "free will,"

The classic example of inductive reasoning is: "All crows we have seen are black; therefore, we believe all crows are black." This *assumes* that had we chosen to look at a different set of crows, we would also find them black. Strictly speaking, every not-looked-at crow might be green. Inductive reasoning assumes that the crows we *chose* to look at were representative of all crows. It assumes we could have *chosen* to look at any other set of crows. Induction and free will are closely related.

Inductive reasoning, which is going from particular cases to a general conclusion, has a logical problem: The only argument for accepting its validity is that it has worked (in particular cases) in the past. But that's inductive reasoning! It's long recognized that the only argument for the validity of inductive reasoning *assumes* what is to be established, and is thus logically not valid. Nevertheless, all science is based on induction. From specific examples, we formulate general rules. We also run our lives and our societies by inductive reasoning. (For example, if I did not eat lunch, I'd be hungry. Or, if he did not pull the trigger, he'd not be sent to jail. We accept such statements as valid because they worked in the past.)

Inductive reasoning entered our box-pairs experiment when we assumed the particular set of box pairs with which we *chose* to do *either* a look-in-the-box experiment *or* an interference experiment was representative of *all* such box pairs. We assumed we could choose to demonstrate either of two contradictory situations. The enigma arose because we assumed that we could have *chosen* to do other than what we in fact did. We assumed we had free will, that our choice was not predetermined by what was "actually" in each set of box pairs.

In our Alice-and-Bob story, and in the actual twin-state photon experiments, the induction assumption implies that the photons observed with a particular polarizer angle were representative of all the photons in the experiment. For example, we implicitly assumed that Alice and Bob (or Clauser or Aspect) could have freely chosen to do Experiment IV with the photons with which they in fact did Experiment II. And if they had done so, they would see the same violation of Bell's inequality. We assume that we do not have a conspiratorial world, one in which the supposedly free choices of the Bell's theorem experimenters were correlated with particular photons.

The role of the free will of the experimenter is usually ignored because of its apparent obviousness. It is, however, a fundamental, though unprovable, assumption in any scientific investigation in which one seeks a general explanation for specific experimental results. (We note that we know of our free will only by our *conscious experience* of it as being free.)

One can evade the quantum enigma by denying it meaningful to even *consider* experiments that were not in fact done, and claim that the conscious perception that we *could* have done them is meaningless. Such denial of free will goes beyond the notion that what we choose to do is determined by the electrochemistry of our brain. The required denial implies a *completely* deterministic and conspiratorial world, one in which our supposedly free choices are programmed to coincide with an external physical situation. Were that true, it would be meaningless to talk of what we *could have* chosen to do. This stance for evading the quantum enigma is a denial of "counterfactual definiteness."

Bell recognized such a *logical* possibility existed, but he hardly considered it a resolution:

> [Even if the polarizer angles are chosen] by the Swiss national lottery machines, or by elaborate computer programmes, or by apparently free willed physicists, or by some combination of all of these, we cannot be sure that [these angles] are not significantly influenced by the same factors that influence [the measurement results]. But this way of arranging quantum mechanical correlations would be even more mind boggling than one in which causal chains go faster than light. Apparently separate parts of the world would be conspiratorially entangled, and our apparent free will would be entangled with them.

Is It Einstein for Whom the Bell Tolls?

Both Einstein and Bohr died before Bell presented his theorem. Surely Bohr would have predicted the experimental result confirming quantum theory. It's not clear what Einstein would have predicted had he seen Bell's proof. He said he believed that quantum theory's predictions would always be correct. But how would he feel if the predicted result was an actual demonstration of what he had ridiculed as "spooky actions"? Would he still insist that separated objects cannot influence each other by faster-than-light connectivities?

Bell, Clauser, and Aspect showed Bohr to be right and Einstein wrong in the EPR argument. But Einstein was right that there was something to be troubled about. It was Einstein who brought quantum theory's full weirdness up front. It was Einstein's objections that stimulated Bell's work and that continue to resonate in today's attempts to come to terms with the strange worldview quantum mechanics forces upon us.

According to Bell:

In his arguments with Bohr, Einstein was wrong in all the details. Bohr understood the actual manipulation of quantum mechanics much better than Einstein. But still, in his philosophy of physics and his idea of what it is all about and what we are doing and should do, Einstein seems to be absolutely admirable. ... [T]here is no doubt that he is, for me, the model of how one should think about physics.

14

Experimental Metaphysics

*All men suppose that what is called wisdom deals
with the first causes and the principles of things.*

—Aristotle, in *Metaphysics*

Metaphysics, literally, "after physics," is the title a first-century editor gave to a collection of Aristotle's philosophical works that came after his book *Physics*. Were Aristotle around today, he would surely explore "first causes" by trying to understand what quantum mechanics is telling us about the world, and about us.

Our title for this chapter, "Experimental Metaphysics," was inspired by a recent collection of essays by that name discussing laboratory experiments exploring the foundations of quantum mechanics. The book's first chapter (by John Clauser) has the provocative title "De Broglie Wave Interference of Small Rocks and Live Viruses," which are the experiments Clauser is proposing.

Because the microscopic realm of atoms differs by so many orders of magnitude from the macroscopic realm of humans, some argue that quantum mechanics has little implication for our human-scale view of Nature, "what's really going on." That was *not* the attitude of Einstein, Bohr, Schrödinger, Heisenberg, and the other developers of quantum theory. In later years, however, as the quantum enigma remained unresolved, and the theory worked so well for all practical purposes, the early concerns waned. That's changed. There's lots of agreement today that we fundamentally don't understand what's going on. At least there's lots of *dis*agreement about what's going on, which is pretty much the same thing.

Bell's theorem and the experiments it fostered are responsible. They did more than confirm the weird predictions of quantum theory. The experiments showed that no future theory could *ever* explain our actual world as a "reasonable" one. Any correct future theory must describe a world in which objects do not have properties that are separately their own, independent of their "observation." In principle, that applies to *all* objects. Even to us?

From a classical physics point of view, some argue that we are just objects governed by biology and chemistry, and therefore ultimately by deterministic physics. However, since Bell's theorem, the human element, free choice, for example, is seen as an issue in fundamental physics questions.

While the free choice of the experimenter was implicit in classical physics there is no classical physics experiment where free choice, a human element, becomes problematic. Although it may never be practical to do a quantum experiment critically involving free choice, a suggested one discussed below comes close.

In the rest of this chapter, we touch on several experiments and proposed experiments that ever more tightly, but mysteriously, connect the strange microscopic world with the "reasonable" macroscopic world we experience.

Macroscopic Realizations

So far, in our telling of an object's existence in two places at once, or its entanglement with another object, the objects were photons, electrons, or atoms, objects small enough to be physically isolated from their macroscopic surroundings. In recent years, quantum phenomena have been extended to larger objects, and even more significantly, to objects with substantial contact with the macroscopic environment. By the time this book is in print there will surely be dramatic phenomena we would have included.

Here's an early example of "two places at once" with an almost macroscopic object. In 1997, researchers at MIT put a clump of *several million* sodium atoms at low temperature in a quantum state called a Bose-Einstein condensate. They then put this *single clump* two places at once separated by a distance larger than a human hair. That's a small separation, but it's a macroscopically seeable one. The *whole clump* was in both places. Each of its atoms was in both places. To demonstrate that this clump, this almost macroscopic object, was in two places at the same time, they did what one always does to demonstrate such a superposition state. They brought the clump from the two regions together to overlap and produce an interference pattern.

Physicists at the University of California, Santa Barbara in 2009 demonstrated a quantum entanglement between two objects big enough to see with your naked eye. Figure 14.1 is an electronic circuit chip made of aluminum in contact with a solid substrate. Each side of the largest white box is 6 mm, a quarter inch. The small white squares on the gray background are superconducting loops, and a current can flow within each of them. A pulse of microwaves directed at the chip entangles the two current flows. Classically, the direction of current flow in the two loops should be completely independent of each other. But after the entangling microwave pulse, the currents flowed in opposite directions, something explained only by the quantum entanglement of these directly seeable objects. Entanglements of circuits like this are the probable basis of quantum computers.

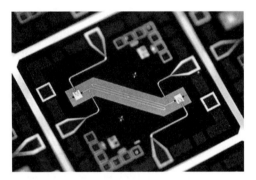

Figure 14.1 The small white squares on the gray background are two macroscopic entangled objects

Scientists at the U.S. National Institute of Science and Technology, in 2008, displayed the first device on a chip that could reasonably be described as a "quantum computer." It even looks a bit like an early computer circuit. Here, trapped ions and associated circuitry can perform at least 160 different computer operations, albeit with only ninety-four percent accuracy. For any practical use, accuracy would need to be greatly improved, and a practical quantum computer would have to link many such devices, by quantum entanglement, Einstein's "spooky actions." In 2009, *Physics World* picked this quantum achievement as the "Breakthrough of the Year."

A March 2010 article in *Nature News* is titled "Scientists Supersize Quantum Mechanics: Largest Ever Object Put into Quantum State." The object was a metal paddle only a thousandth of a millimeter long, but visible to the naked eye in the same way you can see a tiny dust mote in a sunbeam. The little cantilever was cooled to an extremely low temperature until it reached the most motionless state permitted by quantum mechanics, essentially standing still. It was then "excited" to be in a superposition of that motionless state and *simultaneously* in a vibrating state. The paddle was moving and not moving at the same time. (Shades of a cat being dead and alive at the same time!) Even more impressive than the *existence* of this almost macroscopic superposition state is the fact that the paddle was *not* physically isolated. Its base was solidly connected to a block of silicon, which was in physical contact with the experimental apparatus, and ultimately with the rest of the world. It was enough to "isolate" the particular vibration motion, and not necessary to isolate the physical object. It was often considered that *any* contact with the macroscopic surroundings would rapidly collapse a strange superposition. The entanglement of modes of behavior of objects too big to isolate now looks much more feasible. This feat by scientists at the University of California at Santa Barbara was named the 2010 "Breakthrough of the Year" by *Science* magazine. Even before the year was over! It came too late for us to include a picture of the paddle in our book, but you can see it at:

http://www.nature.com/news/2010/100317/full/news.2010.130.html.

In 2011, an article in *Nature* reported the cooperative effort of scientists at five different laboratories to display interference with large organic molecules. The largest contained 430 atoms. This set a new record for putting individual objects in two places at the same time. Moreover, the fact that the molecules had internal temperatures of several hundred degrees Centigrade demonstrated that positional wavefunctions are not necessarily decohered by coupling to internal thermal motions. This makes the apparent display of quantum phenomena in biological systems ever more reasonable. The philosophical significance of their work was not ignored by the authors, who refer to their molecules as "the fattest Schrödinger cats realized to date."

Macroscopic Proposals

Proposals abound for the entanglement or for putting essentially macroscopic objects in two places at once. In some cases there's a further end in mind, such as the sensitive detection of gravity waves. Often the motivation is to display the strangeness of quantum theory on an ever more provocative level.

In 2003 a paper titled "Towards Quantum Superpositions of a Mirror" scientists at the University of Oxford and the University of California, Santa Barbara, claim the result implied by the paper's title, a mirror in a quantum superposition state, "is within reach using a combination of state-of-the-art technologies." The mirror they're talking of is tiny, but it's one seeable with the naked eye. It would be mounted on a tiny lever terminating one arm of an interferometer. A quantum superposition would be indicated by the disappearance of interference and its return as the mirror goes into a superposition state and then returns to its initial state. Experiments in 2006 testing the feasibility of the earlier proposal conclude that it is feasible, though barely, with today's technology.

In 2008, calculations by physicists at the Max-Planck-Institut für Gravitationsphysik at Leibniz and at Potsdam argued that the entanglement of two "heavily macroscopic mirrors" will be achievable within the next decade. The mirrors they analyze are each on the perpendicular arms of an interferometer built to detect gravitational radiation, something predicted by general relativity, but yet to be observed. Gravitational wave interferometers, for which quantum phenomena are proposed, are currently in operation and use mirrors ranging from a few grams up to 40 kg.

A 2008 article in the American Physical Society's *Physical Review Focus*, which publicizes significant physics of wide interest, is titled "Schrödinger's Drum." The allusion, of course, is to Schrödinger's cat. Here the "cat" is an essentially macroscopic one-millimeter-square membrane of silicon nitride that is free to vibrate like a drum and is cooled to a very low quantum state. Researchers at several institutions are discussing such a drum. In a particularly interesting display, a pair of such membranes would be entangled so that an observation of one instantaneously influenced the other–without any physical force connecting them.

Quantum Phenomena in Biology?

The question mark in this section's title reflects our prejudice as physicists that contact with the warm, wet biological environment would frustrate any quantum superposition or entanglement. Countering that concern, perhaps a single aspect of a biological system could be sufficiently decoupled from the rest of the body. An example of such decoupling was demonstrated for the little visible paddle described above. That paddle had to be at extremely low temperature so that vibrating atoms did not disturb the superposition state, a usual requirement for a quantum effect in a many-atom object. Low temperatures would preclude any biological process. But conceivably, there could be a decoupling from the thermal motion. A warm violin string vibrating many thousands of cycles would be a classical analogy. A quantum entanglement in a warm, wet biological environment is hard to believe, but is it less counterintuitive than the quantum enigma itself?

A proposed quantum phenomenon *with* a biological organism, not just *in* a biological process, can raise philosophical issues. In 2009 scientists at the Max-Planck-Institut in Garching and the Institut de Ciències Fotòniques in Barcelona proposed putting living organisms in quantum superposition states, in two places at the same time. They intend to optically levitate an influenza virus, put it in a superposition state by using a light pulse, and subsequently detect the superposition state by reflected light. Their analysis argues for the feasibility of their proposal for even larger living organisms, in particular, tardigrades, or "water bears," which can survive at the low temperatures and in the vacuum required for these experiments. They consider their work "to be a starting point to experimentally address fundamental questions, such as the role of life and consciousness in quantum mechanics."

Explaining the remarkable efficiency of photosynthesis by quantum coherence is not a new idea. But in 2010, chemists at the University of Toronto offered experimental evidence that algae use quantum coherence to harvest light. In photosynthesis, special proteins absorb incoming photons to excite electrons to higher energy to start a series of electron transfers to "photosystems," where the energy of the electrons starts the creation of carbohydrates. Classically, the electrons would find their way to photosystems by random hops. But the high efficiency displayed suggests that electron probability waves sample many paths simultaneously and collapse to find the best ones. To display this, they excited proteins with a laser pulse and used a second laser pulse to see where the electrons went.

Analysis by researchers at the University of Geneva and at the University of Bristol in 2009 show that quantum experiments establishing a violation of Bell's inequality are possible with human eyes as the detectors at one site. Since the human eye cannot reliably detect a single photon, one of the twin-state photons is amplified by cloning it by stimulated emission. What is claimed here is not only that there can be entanglement between two microscopic systems, but also that there can be entanglement between a microscopic object and a macroscopic human system. This can supposedly be so, even in the presence of photon loss to the environment, which might have been expected to wash out the entanglement.

A 2009 article in the *Proceedings of the National Academy of Science* is titled "Some Quantum Weirdness in Physiology." The paper notes that "most modern biomolecular scientists view quantum mechanics much as deists view their God; it merely sets the stage for action and then classically understandable, largely deterministic pictures take over." The paper then comments on a dozen, mostly recent, studies *denying* that mainstream view. These papers report evidence for quantum coherence effects, that is, superpositions and entanglements, in biological systems, principally photosynthesis and vision.

Two other proposals for even weirder quantum phenomena in a biological system, namely, the human brain, one by Roger Penrose and another by Henry Stapp, are treated in chapter 17. Both focus on the issue of consciousness.

Beyond Conventional Wisdom

Any two objects that interact become entangled. After that, whatever happens to one instantaneously influences the other no matter how far apart they are. This has been extensively demonstrated with pairs of microscopic particles, and even with almost macroscopic devices. As entangled objects entangle with yet other objects, the entanglement becomes complex. After interaction with a macroscopic object, any entanglement completely washes out, for all practical purposes. In a sense, however, since *everything* has at least indirectly interacted, there is thus, in principle, a universal connectedness. It's been claimed that you are quantum mechanically entangled with anyone you have ever encountered, presumably more so for more intense encounters. This is, of course, a stretch vastly beyond anything demonstrable, therefore beyond anything meaningful. Complex entanglement becomes essentially *no* entanglement.

However, recent studies suggest entanglement persisting longer than the conventional calculations imply. As one example, in the magnetically sensitive molecule that birds may possibly use as a compass, electrons remain entangled ten to one hundred times longer than expected. Today's arguments may be too pessimistic as to the survival of quantum effects. The flawed history of the theoretical limits of quantum effects suggests open mindedness.

For example, quantum information theorist Seth Lloyd in 2008 found "quantum benefits" to survive even *after* entanglement decoheres. It is claimed that this unexpected effect may make it possible to more accurately view an object by illuminating it with photons in a twin state. One could store one photon of each pair, compare it with its reflected twin, and reject any stray photons, which would not have twins in storage. This might work even though the entanglement of the returning photons with their twins decohered as a result of their interaction with the object they reflected from. Lloyd found, to his surprise, that "to gain full enhancement of quantum illumination, all entanglement must be destroyed." Other physicists were skeptical, until they checked and confirmed Lloyd's calculations.

When an object in a superposition state with *no* particular position encounters a macroscopic object, its superposition wavefunction collapses to a *particular* position. Similarly, when a photon encounters a vertical polarizer followed by a Geiger counter, the firing or not-firing of the Geiger indicates whether the photon was vertical or horizontal. That's the Copenhagen interpretation, and it is surely true, for all practical purposes.

How do we know it's *actually* true? How can we be *sure* that the Geiger counter does not go into a superposition state of being simultaneously both fired and not-fired before actually being observed by a human? That's a silly question. But it's not one with an *experimentally* established answer.

Here's a related question: How can we be *absolutely* sure that in an actual version of our Bell's theorem experiment that Alice's supposedly free choice of her polarizer angle is truly independent of Bob's supposedly free choice of his polarizer angle? Bell's theorem and the experiments establishing a violation of Bell's inequality crucially depend on that assumption.

In the actual laboratory experiments, only several meters could separate Alice and Bob. To be absolutely sure that it was not physically possible for a choice of polarization angle at Alice's site to affect the choice at Bob's site, the two choices would have to be made closer in time than it would take light to go the several meters from Alice to Bob. They would have to be made within a fraction of a microsecond.

Humans can't make choices that fast. In the actual experiments, fast electronic devices made the "Alice" and "Bob" choices. Can we be absolutely sure that choices made by boxes of electronics are truly independent? It can't be completely ruled out that *something* in the past history of these devices correlated the two decisions. Extremely unlikely, hard to believe. But the interpretation these experiments demand is also hard to believe.

The choices we are most sure of being independently made are our own conscious free choices. And we grant such free will to fellow humans, Alice and Bob. Ideally, therefore, we would want the EPR-Bell experiment done with humans, rather than electronic devices, choosing the polarization angles. However, we must be *absolutely* sure that during the second or so that it takes Alice to make a choice, she in no way communicates her choice to Bob. In an ideal experiment, we would want to establish that what Alice chooses to observe could not possibly affect what Bob chooses to observe.

To exclude any possible communication between human observers as detectors of photons, Anthony Leggett (Physics Nobel Prize, 2003) suggested that observers could be separated by the distance light (or any physical interaction) can travel in a second. That's 186,000 miles. That's a long distance, but less than the 250,000 miles to the moon. We can separate humans in space by that distance and do a Bell's theorem experiment between them. "That will happen someday. There is no doubt in my mind" says Anton Zeilinger, in whose lab buckyballs were put in two places at once.

We're not at ease with non-physical "influences." Or with reality creation by "observation." Certainly not with *history* creation. Experimental metaphysics may some day lead to explanations beyond today's quantum theory. But Zeilinger warns us: "This new theory will be so much stranger . . . people attacking quantum mechanics now will long to have it back." We earlier quoted John Bell telling us that we are likely to be "astonished."

15

What's Going On?
Interpreting the Quantum Enigma

*You know something's happening here, but you
don't know what it is.*

—Bob Dylan

*It is a striking fact that almost all the
interpretations of quantum mechanics…depend to
some degree on the presence of consciousness for
providing the "observer" that is required for…the
emergence of a classical-like world.*

—Roger Penrose

Physicists and Consciousness

More physicists today are willing to face up to the quantum enigma, and
some struggle to *interpret* what quantum mechanics might be telling us.
Several interpretations today contend with the Copenhagen. Before we tell
of several of them, we reflect on how physicists can approach the enigma.

To their dying days, Bohr and Einstein disagreed about quantum
theory. For Bohr, the theory with its Copenhagen interpretation was the
proper basis for physics. Einstein rejected Copenhagen's concept of a
physical reality created by "observation." He nevertheless accepted a *goal*
of the Copenhagen interpretation, which was to allow physics to move on
without dealing with consciousness. Most physicists (including ourselves)
would agree that consciousness *itself* is beyond the physics discipline, not
something to be studied in a physics department.

It's not that physicists are averse to ranging widely. For example, a famous mathematical treatment of predator-prey relations (foxes and rabbits isolated on an island) was published in *Reviews of Modern Physics*. On Wall Street, physicists model arbitrage (and are called "quants"). One of us (Bruce) has strayed into biology to analyze how animals detect the Earth's magnetic field. Such things are happily accepted as part of the physics discipline, while the study of consciousness is not. Here is a working definition of physics that might make that attitude understandable: Physics is the study of those natural phenomena that are successfully treatable with well-specified and testable models.

For example, physics treats atoms and *simple* molecules. Chemistry, on the other hand, deals with *all* molecules, most of whose electron distributions are too complex to be well-specified. A physicist might study a readily characterized biological system, but the functioning of a complex organism is the domain of biologists.

Anything not successfully treatable with a well-specified and testable model is rather quickly *defined* out of physics. When we focus on consciousness in chapter 16, we offer no such model. There is none. Until one is developed, consciousness won't qualify for study as physics.

This is reason enough for not studying consciousness in physics departments, but it hardly explains the emotion that talking of our discipline's *encounter* with consciousness can arouse. Recently, I (Fred) gave a talk in our physics department reporting on a conference I attended at Princeton University honoring quantum cosmologist John Wheeler on his ninetieth birthday. Several talks on cosmology and the foundations of quantum mechanics referred to consciousness. When I reported in our department on this conference and on our related interests, I was heckled by two older faculty: "You guys are taking physics back to the Dark Ages!" And: "Spend your time doing good physics, not this nonsense!" Physics graduate students in the audience, on the other hand, seemed fascinated.

Classical physics, with its mechanical picture of the world, has been used to deny the existence of anything beyond the strictly mechanistic. *Quantum physics denies that denial.* It hints of something beyond what we usually consider physics, beyond what we *usually* consider the "physical world." *But that's the extent of it!* We should be careful. In dealing with the mysteries of quantum mechanics, we walk the edge of a slippery slope.

A recent movie, strangely titled *What the #$*! Do We (K)now!?* (It's informally called *What the Bleep?*), is described by *Time* magazine as "an odd hybrid of science documentary and spiritual revelation featuring a Greek chorus of Ph.D.s and mystics talking about quantum physics." The movie uses special effects to display quantum phenomena with macroscopic objects. For example, it greatly exaggerates the quantum uncertainty in the position of a basketball. That's easily understood as pedagogical hyperbole. The movie's allusion to quantum mechanics encountering the realm of consciousness is also valid. But the movie then ventures into "quantum insights" leading a woman to toss away her antidepressant medication, to the "quantum channeling" of a 35,000-year-old Atlantis god, and on to yet greater nonsense.

What's in the minds of some people leaving the theater? If it's that physicists spend their time dealing with the "spiritual revelations" the movie describes, we're embarrassed. If viewers think the physicists in the movie expressing these mystical ideas represent more than the tiniest fraction of the physics community, they've been misled. The movie slides far down the slippery slope.

Our antidote for sensationalistic and misleading treatments of the implications of quantum mechanics would be for the physics discipline to be more open to the discussion of the quantum enigma, particularly in conceptual physics courses. Keeping our skeleton in the closet concedes the field to the purveyors of pseudoscience.

Why Interpretations?

Should a trustworthy friend tell you something that seems ridiculous, you'd try to *interpret* what he or she might really mean. Trustworthy physics demonstrations tell us something that seems ridiculous. We therefore try to *interpret* what those results might really mean. While there is complete consensus on the experimental results, there is no consensus on their meaning. Many interpretations currently contend. Every one of them displays a quantum weirdness. The Copenhagen interpretation provides a way for physicists to ignore the weirdness, for all *practical* purposes at least, and get on with the business of physics. Appropriately, most physicists accept that. But it's also worth exploring what Nature is telling us.

As John Bell says (only a little bit tongue-in-cheek):

> Is it not good to know what follows from what, even if it is not necessary FAPP [for all practical purposes]? Suppose for example that quantum mechanics were found to *resist* precise formulation. Suppose that when formulation beyond FAPP is attempted, we find an unmovable finger obstinately pointing outside the subject, to the mind of the observer, to the Hindu scriptures, to God, or even only Gravitation? Would that not be very, very interesting?

Going beyond FAPP to interpret quantum theory is today a growth industry, and a contentious field, though a very small fraction of all physicists are involved. Each of the interpretations currently proposed looks differently at what quantum mechanics reveals about our world. At times, different interpretations seem to say the same thing in different terms. Or two interpretations might even contradict each other. That's OK. While scientific *theories* must be testable, *interpretations* need not be. All "interpretations" assume the same experimental facts.

Most interpretations tacitly accept that quantum mechanics ultimately *encounters* the problem of conscious observation. However, they usually start out with the *presumption* that physicists should deal with a physical world independent of the human observer.

Murray Gell-Mann, for example, begins a popular treatment of quantum physics by saying: "[T]he universe presumably couldn't care less whether human beings evolved on some obscure planet to study its history; it goes on obeying the quantum-mechanical laws of physics irrespective of observation by physicists." In talking about *classical* physics, Gell-Mann's explicit *presumption* that the laws of physics are independent of the human observer would be assumed *without* his saying it.

Every interpretation presents a weird worldview. How could it be otherwise? We saw the weirdness of quantum mechanics right up front in the *theory-neutral* experimental facts. An interpretation of those facts that goes beyond "Shut up and calculate!" *must* be weird.

Though the interpretations we will discuss have been developed with extensive mathematical and logical analysis, we present each of them in a few non-technical paragraphs. We treat today's three main alternatives to

Copenhagen (Decoherence, Many Worlds, and Bohm) in a bit more detail. Substantial understanding any of these is *not* crucial for what follows. We just wish to show different perspectives on what the facts of quantum experiments might imply. Consider how each interpretation has physics inevitably encountering consciousness, but also how each avoids requiring any serious relationship of physics with consciousness. (Apologies to anyone whose favorite interpretation we have left out.)

Ten Currently Contending Interpretations

Copenhagen

The Copenhagen interpretation, physics' orthodox stance, is the way physicists, including ourselves, teach and use quantum theory. We say little about it here since we devoted all of chapter 10 to it. In the standard version of Copenhagen, observation creates the physical reality of the microscopic world, but the "observer" can, for all *practical* purposes, be considered to be a macroscopic measuring device, a Geiger counter, for example.

Copenhagen addresses the quantum enigma by telling us to pragmatically use quantum physics for the microworld, and classical physics for the macroworld. Since we supposedly never see the microworld "directly," we can just ignore its weirdness and thus ignore physics' encounter with consciousness. Since quantum weirdness is today seen with larger and larger objects, the ignoring gets harder, and other interpretations proliferate.

Extreme Copenhagen

Aage Bohr (a son of Niels Bohr, and also a Nobel laureate in physics) and Ole Ulfbeck hold that the Copenhagen interpretation does not go far enough. The standard Copenhagen interpretation allows physics to ignore its encounter with consciousness by considering observer-created reality confined to the microscopic world. Bohr and Ulfbeck explicitly deny the *existence* of the microworld. In this view, there are no atoms.

Bohr and Ulfbeck intend their outlook to apply generally, but discuss it in terms of the clicks of a Geiger counter and the correlated changes in a piece of uranium. We normally consider uranium nuclei to randomly emit alpha particles (helium nuclei) and thereby become thorium nuclei. In the Copenhagen interpretation, the extended wavefunction of the alpha particle is, for all practical purposes, collapsed by the Geiger counter to the position at which the counter observed it.

Bohr and Ulfbeck find such a for-all-practical-purposes resolution unacceptable. Taking the bull by the horns, they claim that atomic-scale objects do not exist at all. Nothing moved through the space between the piece of changed uranium and the clicking Geiger counter. The consciously experienced clicks in counters are "genuinely fortuitous" events correlated with changes in a remote piece of uranium without the intermediary of alpha particles.

As they say it:

> The notion of particles as objects in space, taken over from classical physics, is thereby eliminated. . . . The click being genuinely fortuitous, is no longer produced by a particle entering the counter, as has been a foregone conclusion in quantum mechanics. . . . The downward path from macroscopic events in spacetime, which in standard quantum mechanics continues into the region of particles, does not extend beyond the onset of clicks.

Accordingly, when chemists, biologists, and engineers talk of photons, electrons, atoms, and molecules, they are merely dealing with models without physical reality. No photons pass through the space between the light bulb and your eye. No air molecules bounce off the canvas sheet to push the sailboat through the water.

Decoherence and Decoherent Histories

Some years ago, a physicist would use the word "collapse" to describe the process of observation by which a superposition state wavefunction becomes an observed single reality. Instead of "collapse," a physicist today might use the word "decoherence." It refers to the now well-studied process by which the wavefunction of a microscopic object interacts with the macroscopic environment to produce the result we actually observe, what Copenhagen accounts for with the unexplained "collapse" of the wavefunction. Decoherence can be considered an extension of Copenhagen.

Let's go to our box-pair example. Consider an atom whose wavefunction is simultaneously in two boxes. We now send a photon through transparent windows in one of the boxes. Were the atom in that box, the photon would bounce off the atom in a new direction. Were the atom in the other box, it would go straight through unchanged in direction. Since the atom is actually simultaneously in both boxes, the photon does *both* things. The atom's wavefunction becomes entangled with that of the photon. Each photon *randomly* disrupts the phase relation between the parts of each atom's wavefunction in the two boxes. The parts of the wavefunction out of each box pair then cancel at different places on the detection screen. Thus no pattern forms.

Thus, atoms whose wavefunction have been entangled with those of photons cannot participate in an interference pattern. The phase of those atoms would be scrambled, or "decohere." The atoms would land in an essentially uniform distribution. Without an interference pattern, there can be no evidence of atoms having been simultaneously in both boxes.

Actually, if the photons in question do not interact with other objects, a tricky two-body interference experiment with the set of box pairs *and* the photons could demonstrate that each atom had indeed been simultaneously in *both* boxes, and that each photon had *both* bounced off an atom *and* gone through an empty box.

Suppose, however, that the photons pass through our boxes and then encounter the macroscopic environment. Assuming thermal randomness, one can calculate the extremely short time after which an interference experiment becomes impossible, for all practical purposes. After that, one cannot display a quantum enigma. Averaging over the decohered wavefunctions of the atoms leaves us with an equation for a classical-like probability for each atom actually existing wholly in one or the other box of its pair. In experimental tests the rate of decoherence for objects as massive as large molecules accurately confirms decoherence theory calculations.

Since no observer, conscious or otherwise, need be mentioned, some argue that this resolves the observer problem. Others see a fundamental non sequitur in that argument. Those classical-*like* probabilities are still probabilities of what will be *observed*. They are *not* true classical probabilities of something that actually *exists*. Decoherence is then merely a FAPP solution to the quantum enigma. W. H. Zurek, a major developer of the decoherence interpretation, acknowledges that consciousness is at least *ultimately* encountered:

> An exhaustive answer to this question [the perception of a unique reality] would undoubtedly have to involve a model of "consciousness," since what we are really asking concerns our (observer's) impression that "we are conscious" of just one of the alternatives.

An extension of the decoherence concept, "decoherent histories," boldly applies quantum theory to the entire universe, from beginning to end. No observers were around in the early universe, and no *external* observers have existed at any time; the universe includes everything. Since one can't deal with the infinite complexity of the universe, one treats only certain aspects and averages over the rest.

For a very rough idea of how this can work, consider an atom on its way to its box pair passing through a thin gas of much lighter atoms. Gently bouncing along, our atom is not strongly deviating from its two paths. But the parts of the wavefunction on each path, changing phase a tiny bit with each bounce, decohere enough that no interference experiment is possible, for all practical purposes. By averaging over the vast number of possible histories, one for every possible series of bounces, we come to two course-grained histories, one for the atom in each box. Now we claim that only one of these two histories is an actual history, and the other is just a history that had been possible.

In their development of this interpretation, Gell-Mann and James Hartle discuss the evolution of an IGUS, an "Information Gathering and Utilizing System." Presumably, the IGUS ultimately becomes an observer with at least the conscious illusion of free will.

Many Worlds

The Many Worlds interpretation accepts what quantum theory says *literally*. Where the Copenhagen interpretation has observation mysteriously collapsing the atom's wavefunction into a single box, and Schrödinger's cat into the living *or* dead state, the Many Worlds interpretation just says "no" to collapse. Quantum theory says the cat is simultaneously alive and dead. So be it! In one world, Schrödinger's cat is alive, and in another it is dead.

Hugh Everett came up with the Many Worlds idea in the 1950s to allow cosmologists to deal with a wavefunction for the entire universe. With no need for "observers" to collapse the wavefunction, the Many Worlds interpretation presumes to resolve the quantum enigma by the sensible-seeming ploy of including consciousness as part of the physical universe described by quantum mechanics.

In the Many Worlds interpretation, you are part of the universal wavefunction. Consider our box pairs. Looking into one of the boxes, you entangle with the atom's superposition state. You go into a superposition state *both* of having seen the atom in the box you looked in and also of having seen that box empty. There are now two of you, one in each of two parallel worlds. The consciousness of each one of you is unaware of the other "you." Instead of looking in a box, yet another "you" did an interference experiment. Nothing we actually experience conflicts with this bizarre view.

To bring more than one observer into the picture, let's go back to Schrödinger's cat. Alice looks in the box while Bob is far away. In one world Alice, call her $Alice_1$, sees a live cat. In the other, $Alice_2$ sees a dead cat. At this point Bob is also in both worlds, but Bob_1 and Bob_2 are essentially identical. Should Bob_1 meet $Alice_1$, he would help her get milk for the hungry cat. Bob_2 would help $Alice_2$ bury the dead cat. Macroscopic objects $Alice_2$ and Bob_1 exist in different worlds and, for all practical purposes, never encounter each other.

After Bell's theorem and the experiments it allowed, we know that our actual world perhaps cannot have reality and, certainly cannot have separability. In the Many Worlds interpretation there is no separability. In that world in which Alice finds the cat alive, Bob instantly becomes a person in the cat-alive world. And there is clearly no *single* reality, which seems equivalent to no reality.

The Many Worlds interpretation stirs strong feelings. One academic author decries it as "profligate" and refers to its proposer as a "chain-smoking, horned-Cadillac-driving, multimillionaire weapons research analyst." (At the time Everett proposed it, he was just a graduate student.) On the other hand, a leader in quantum computing writes that the Many Worlds interpretation "makes more sense in so many ways than any previous world-view, and certainly more than the cynical pragmatism which too often nowadays serves as a surrogate for a world-view among scientists." (By "cynical pragmatism" he surely means the unquestioning acceptance of Copenhagen.)

The Many Worlds interpretation is today the favorite interpretation of many quantum cosmologists for the early universe they consider. They can ignore the observer problem. No observers around then. Since the universe includes everything, it is by definition isolated from any external "environment." Decoherence thus need not be an issue. A quantum cosmologist colleague tells us that the Many Worlds interpretation is his favorite, although he doesn't like it.

There's an unresolved problem with the Many Worlds interpretation: What constitutes an observation? *When* does the world split? The splitting into two worlds is presumably just a way of speaking. Are infinitely many worlds continuously created?

In any event, the Many Worlds interpretation vastly extends what Copernicus started. Not only are we removed from the center of the cosmos to a tiny spot in a limitless universe, but the world we experience is just a minute fraction of all worlds. However, "we" exist in many of them. Many Worlds, the most bizarre description of reality ever seriously proposed, provides a fascinating base for speculation, and for science fiction.

Transactional

The transactional interpretation approaches the intuitive challenges posed by Schrödinger cats and universal connectedness by allowing the wavefunction to evolve backward as well as forward in time. The future thus affects the past. This does, of course, alter the way we look at what's happening.

For instance, here's an example offered by the transactional interpretation's proposer, John Cramer:

> When we stand in the dark and look at a star a hundred light years away, not only have the retarded light waves from the star been traveling for a hundred years to reach our eyes, but the advanced waves generated by absorption processes within our

eyes have reached a hundred years into the past, completing the transaction that permitted the star to shine in our direction.

While this backward-in-time approach still encounters the conscious observer, we do end up with the quantum enigma packaged into what appears to be a single mystery.

Bohm

In 1952 a maverick young physicist, David Bohm, did the "impossible:" He produced a *counterexample* to the long-accepted theorem claiming to show that the experimental facts were inconsistent with hidden variables. Bohm's counterexample reproduced all the predictions of quantum theory with an interpretation having hidden variables, quantities not appearing in the standard formulation of quantum theory. His "hidden variables" were the actual positions of the particles. It was Bohm's work that inspired John Bell to find the mathematical flaw in the no-hidden-variables proof and eventually produce Bell's theorem.

Bohm was also a maverick politically. After he refused to testify before the House Un-American Activities Committee, Princeton University fired him, and he could not get another academic job in the United States.

Bohm starts his interpretation by assuming that his particles, on average, initially have the distribution demanded by the Schrödinger equation. Then, with straightforward mathematics, he deduces a "quantum force" that acts on his particles to make them continue to obey the Schrödinger equation. The quantum force is generally referred to in terms of a "quantum potential."

The quantum potential guides rather than pushes. Bohm offers the analogy of the radio beacon directing a ship. The universal connectedness intrinsic to quantum theory appears right up front in this interpretation. The quantum potential that an object experiences depends on the instantaneous positions of all the objects the one in question has ever interacted with, and with all the objects *those* objects had ever interacted with. In principle, that includes an interaction with everything in the universe. Bohm's quantum potential provides Bohr's "influences," what Einstein called "spooky actions."

The Bohm interpretation describes a physically real, completely deterministic world. The universal instantaneous quantum potential demands a "super-deterministic" world. Quantum randomness appears only because we cannot know the precise initial position and velocity of each particle. There is no unexplained wavefunction collapse, as there is in the Copenhagen interpretation; there is no unexplained splitting of consciousnesses as there is in the Many Worlds interpretation. Some claim the Bohm interpretation resolves the observer problem of quantum mechanics, or at least makes it a benign problem, as it is in Newtonian physics.

Others, including Bohm, see it differently. Unlike the Newtonian atom that merely enters a single box of a pair, the Bohmian atom entering a single box also "knows" the position of the other box. Through the quantum potential, the macroscopic box pair has always been in instantaneous communication with the rest of the world, and thus with the macroscopic device that earlier released the atom, and thus with the incident atom. The quantum potential connects all of this from the beginning, and thus even determines where the atom would land in any later interference pattern. The human who arranged the experiment, presumably also a physical object, influences the quantum potential as well. (And is influenced by it?)

As in the Many Worlds interpretation, since there is no collapse, the part of the wavefunction corresponding to what was not in fact seen continues on forever: We may find Schrödinger's cat alive, but the part of the wavefunction containing the possibility of the dead cat, and its owner burying it, goes on. We may ignore this part of the wavefunction, for all practical purposes, since it has entangled with the environment. But in this interpretation it is real and, in principle at least, has future consequences.

Bohm accepted physics' encounter with consciousness. In their highly technical 1993 book on quantum theory, *The Undivided Universe*, whose title emphasizes that quantum theory applies to the macroscopic as well as the microscopic, Bohm and Basil Hiley write:

> Throughout this book it has been our position that the quantum theory itself can be understood without bringing in consciousness and that as far as research in physics is concerned, at least in the present general period, this is probably the best approach. However, the intuition that consciousness and quantum theory are in some sense related seems to be a good one.

That evening when Einstein tried to tell me (Bruce) and a fellow physics graduate student about his problems with quantum mechanics, he also remarked: "David [Bohm] did something good, but it was not what I told him." Having never been exposed to these problems in our studies of quantum mechanics, we did not know what Einstein was talking about. I wish I had been able to ask what it was that he told Bohm.

Ithaca

David Mermin of Cornell University in Ithaca, New York, proposing what he calls the "Ithaca interpretation," identifies two "major puzzles": objective probability, which arises only in quantum theory, and the phenomenon of consciousness.

Classical probability is subjective, a measure of one's ignorance. *Quantum* probability is objective, the same for everyone. For the atom in a box pair, the quantum probability is not a measure of someone's uncertainty of what *is* but the likelihood of what anyone would *observe*. The Ithaca interpretation takes objective probability as a *primitive* concept, one incapable of further reduction. Ithaca reduces the mysteries of quantum mechanics to this single puzzle.

According to Ithaca, quantum mechanics is telling us that "correlations have physical reality; that which they correlate do not." For example, unobserved twin-state photons have no *particular* polarization, but they have the *same* polarization. Only the *correlation* of their polarizations is a physical reality; the polarizations themselves are not. Or, for example, if the positions of two atoms are entangled, only their separation is a reality, while the position of each atom is not.

What if, for example, we observe a photon's polarization with a macroscopic apparatus whose scale would read differently for two states of a photon's polarization? If we consider the apparatus quantum mechanically, it merely becomes correlated with the photon's polarization. According to quantum theory, the scale should read both ways. But we always see it read one way *or* the other.

Here's how Mermin deals with this in the Ithaca interpretation:

> When *I* look at the scale of the apparatus *I know* what it reads.
> Those absurdly delicate, hopelessly inaccessible, global system

correlations *obviously* vanish completely when they connect up with *me*. Whether this is because consciousness is beyond the range of phenomena that quantum mechanics is capable of dealing with, or because it has infinitely many degrees of freedom or special super-selection rules of its own, I would not presume to guess. But this is a puzzle about consciousness which should not get mixed up with efforts to understand quantum mechanics as a theory of subsystem correlations in the nonconscious world. (Emphasis in original.)

Ithaca steps aside from physics' encounter with consciousness to confine the quantum enigma to the problem of objective probability. The encounter with consciousness is not denied. Ithaca assigns consciousness to a "reality" larger than the "physical reality" to which physics, for the present at least, should be restricted. This modest interpretation of the quantum enigma just admits a mystery.

Quantum Information

An interpretation that has gained favor among those studying quantum computing might be called the "quantum information interpretation." It holds that the wavefunction represents only *information* about possible measurements on a physical system. The wavefunction is now not to be identified with the actual physical system. It does not even *describe* the physical system under consideration.

In this interpretation the wavefunction, or the quantum state, provides only a compact mathematical device for calculating the correlations between observations, for predicting the result of a subsequent measurement from an initial measurement. The quantum state is thus not an objective physical thing; *it's only knowledge*. This interpretation can be seen as a blend of the Ithaca interpretation with its focus on correlations and a version of Copenhagen in which Bohr tells that the purpose of physical law is "only to track down, so far as it is possible, relations between the manifold aspects of our experience."

The quantum information interpretation evades the encounter with consciousness by limiting the quantum state to being only the *knowledge* of possible *observations*. In some sense, it therefore limits the scope of quantum theory to being *only* about consciousness.

Quantum Logic

It's by considering the experiments we *might* have done, but in fact did *not* do, that the quantum enigma arises. Quantum logic denies that it is meaningful to *consider* actions that were not in fact done. It denies counterfactual definiteness. Quantum logic "resolves" the enigma by revising the rules of logic to fit quantum theory.

Quantum logic is an intriguing intellectual exercise and the viewpoint of some quantum theorists. However, since *any conceivable* observations can be "explained" by adopting rules of logic to fit, it hardly provides a comfortable resolution of the quantum measurement problem.

Moreover, in our conscious experience, we must consider alternatives to what we might or might not do. Denying counterfactual definiteness goes beyond merely denying "free will" in the sense that our choices are totally determined by the electrochemistry of our brain. It demands our supposedly free choices to be completely correlated with the external physical situation. We would then be essentially robots in a completely deterministic world. As a resolution of the quantum enigma, this assumption is, to use John Bell's words quoted in chapter 13, "more mind boggling" than the enigma it presumes to resolve.

GRW

To explain why big things are never seen in superposition states, Ghirardi, Rimini, and Weber, with the "GRW interpretation," modify the Schrödinger equation to make wavefunctions randomly collapse every now and then. For things as small as atoms, a collapse occurs only every billion years or so.

Such infrequent collapse would not affect an interference experiment with isolated atoms taking place in a much shorter time. But suppose an atom was in contact with its neighboring atoms in a larger object, say, Schrödinger's cat in a superposition state of alive and dead. That atom would be entangled with its neighbors, and through them entangled with all the other atoms in the cat. That atom's random collapse from its two superposition-state positions in the simultaneously alive and dead cat to a single position characteristic of either the living or dead cat would trigger the collapse of the whole cat to the living *or* dead state. There are so many atoms in a cat that even if a single atom collapsed only every billion years, at least one atom would collapse every micromicrosecond. The cat could thus remain in a superposition of living and dead states only briefly.

Strictly speaking, the GRW scheme is not an *interpretation* of the theory since it proposes a *change* of the theory. It is a change that would allow the macroscopic objects of our perception to be perfectly definite in *principle*, not just for all *practical* purposes. Such a result would satisfy some.

There is as yet no experimental evidence for the GRW phenomenon. Moreover, as experiments with large molecules show the transition to classical-like probabilities following decoherence calculations, the point at which a GRW phenomenon can become effective is pushed to larger and larger objects. That would leave the reality of objects smaller than that, and their experimentally confirmed lack of separability, as an enigma.

Penrose and Stapp Interpretations

Two proposals, one by Roger Penrose and another by Henry Stapp, might be called interpretations but actually include physical speculations involving consciousness. We address these in chapter 17.

What Can Interpretations Accomplish?

Some interpretations of quantum mechanics resolve the measurement problem for all practical purposes. Of course, there never *was* a problem, for all *practical* purposes. The predictions of the theory work perfectly. It's the strange worldview the experimental facts display that makes us ask, "What's going on?" The wide range of today's contending interpretations shows that profound questions about our world (and about us?) are wide open.

Quantum mechanics shows that our reasonable, everyday worldview is fundamentally flawed. Interpretations of what the theory tells us offer different worldviews. But every one of them involves the mysterious intrusion of the conscious observer into the physical world. Is it possible that some yet-to-be proposed interpretation of the theory will resolve the enigma without an encounter with consciousness?

No. The encounter with consciousness arises directly in the *quantum-theory-neutral* experimental demonstration. Therefore, no mere interpretation of the *theory* can avoid the encounter. But every interpretation allows physics to avoid *dealing* with consciousness. Here's how John Wheeler puts the dichotomy:

> Useful as it is under everyday circumstances to say that the world exists "out there" independent of us, that view can no longer be upheld. There is a strange sense in which this is a "participatory universe."

But immediately after stating that, Wheeler cautions:

> "Consciousness" has nothing whatsoever to do with the quantum process. We are dealing with an event that makes itself known by an irreversible act of amplification, by an indelible record, an act of registration. . . . [*Meaning*] is a separate part of the story, important but not to be confused with "quantum phenomenon."

We take this as an injunction to physicists (as *physicists*) to concentrate on the quantum phenomena themselves, not the *meaning* of the phenomena. For all practical purposes, quantum theory needs no interpretation. It works perfectly to predict the results of any particular experiment we choose.

However, some of us, as physicists, or just as wonderers, ponder the meaning and try to understand what's really going on. This has long been an attitude of many eminent physicists (including, at times, Wheeler). It's an attitude that today gains acceptance.

The growth of that acceptance bothers some physicists and stimulates challenges. Moreover, the now increasingly frequent pseudo-scientific treatments of quantum mechanics, like the movie *What the Bleep?*, make physicists squirm and motivate them to minimize the enigma. We physicists tend to keep our skeleton in the closet, and some even deny its existence.

For example, in 1998, an article titled "Quantum Theory without Observers," spanning two issues of *Physics Today*, argued that several interpretations, principally the Bohm interpretation, *eliminate* a role for the observer in quantum mechanics. (Bohm himself, as quoted above, would not agree.) When such arguments are put forth, it is usually unclear whether such elimination of the observer is proposed *in principle* or as merely a solution of the quantum enigma for all *practical* purposes, a FAPPTRAP, to use Bell's put-down of the for-all-practical-purposes argument when it is presumed to resolve fundamental problems. While the attitude of this *Physics Today* article matches the sympathies of perhaps the majority of today's physics community, times are changing.

Eight decades after the Schrödinger equation, the meaning of physics' encounter with consciousness is increasingly in contention. When experts can't agree, you can choose your expert. Or speculate on your own.

"What's going on?" is an open question, and one that motivates us to quote our chapter's opening epigraphs: "You know something's happening here, but you don't know what it is."

Starting with quantum mechanics, we have encountered consciousness. Our next chapter starts with consciousness and approaches the encounter from the other direction.

16

The Mystery of Consciousness

What is meant by consciousness we need not discuss; it is beyond all doubt.

—Sigmund Freud

Consciousness poses the most baffling problems in the science of the mind. There is nothing that we know more intimately than conscious experience, but there is nothing that is harder to explain.

—David Chalmers

Does consciousness collapse wavefunctions? That question, raised at the beginning of quantum theory, cannot be answered. It can't even be well posed. Consciousness itself is a mystery.

When we described the experimentally demonstrated quantum facts and the quantum theory explaining those facts (as distinct from the theory's several contending interpretations), we presented the undisputed consensus of the physics community. We cannot describe such a consensus in our discussion of consciousness. There is none. There is, of course, a large amount of undisputed experimental data, but diametrically opposed explanations of that data are strongly held. We have our own take, but, you may notice, we waver.

Until the 1960s, behaviorist-dominated psychology avoided the term "consciousness" in any discussion that presumed to be scientific. There has since been an explosion of interest in consciousness. Some attribute this to the striking developments in brain imaging technology that allow seeing which parts of the brain become active with particular stimuli. But according to an editor of the *Journal of Consciousness Studies*:

> It is more likely that the re-emergence of consciousness studies occurred for sociological reasons: The students of the 1960s, who enjoyed a rich extra-curricular approach to "consciousness studies" (even if some of them didn't inhale), are now running the science departments.

Interest in the foundations of quantum mechanics grows at the same time as does interest in consciousness. And connections are seriously proposed. There's something in the air.

What Is Consciousness?

We've talked about consciousness but never clearly defined it. Dictionary definitions of "consciousness" are little better than those for "physics." We've been using "consciousness" as roughly equivalent to "awareness." For us, "consciousness" most definitely includes the perception of free choice by the experimenter. This use of "consciousness" is that quite standard in the treatment of the quantum measurement problem. Ultimately, a definition is manifest by the *use* of the term. (As Humpty Dumpty told Alice: "When *I* use a word . . . it means just what I choose it to mean," and the philosopher Wittgenstein, who taught that a word is *defined* by its use, would more or less agree.)

One can know of the existence of consciousness in *no* other way than through our first-person feeling of awareness, or the second-person reports of others. (In our following chapter we suggest an apparent quantum challenge to this limitation.)

We do not discuss many of the things found in treatments of consciousness from a psychological point of view. We do not, for example, talk of optical illusions, mental disturbances, self-consciousness, or Freud's seat of hidden emotions, the *unconscious*. We also don't discuss the many, as yet untestable, theories of consciousness in the current literature that do not impinge on the quantum enigma.

Our concern is with that "consciousness" related to the observer's free choice of experiment, the consciousness that physics encounters.

A somewhat closer connection with psychology and neurology will arise when we soon address Chalmers' "hard problem of consciousness."

Our frequent example of physics' encounter with consciousness is the decision to observe an object in a single box *causing* it to be wholly there. We say "causing" only because the observer presumably *could* have chosen to do an interference observation establishing a contradictory situation—that the object was *not* wholly in a single box. The observer could have, we assume, chosen to establish that the object was a wave simultaneously in two boxes.

Does such a demonstration necessarily require a *conscious* observer? Couldn't a not-conscious mechanical robot, or even a Geiger counter, do the observing? It depends on what you mean by "observing." For now, just recall that if that robot or Geiger counter were isolated from the rest of the world, and was governed by quantum theory, it would merely become entangled as part of a total superposition state, as did Schrödinger's cat. In that sense, it would not *observe*.

The quantum enigma arises from the *assumption* that experimenters can freely choose between two experiments, two experiments that yield *contradictory* results. We assume that the experimenters had the "free will" to make that choice. However, we can't evade the quantum enigma by denying the free will of the experimenters, that is, merely by having their choices somehow determined by the electrochemistry of their brains. To evade the quantum enigma, the required denial of free will must go much further. It must include the denial of counterfactual definiteness. That denial must include the assumption of a "conspiratorial" world. (In our example, the experimenter's "choices" would have to match the physical situation in the box pairs.)

Today's discussions of free will in psychology or neurophysiology usually focus more narrowly on whether the choices we make are somehow predetermined by the electrochemistry of our brain. *This* free will issue is therefore peripheral to the quantum enigma. But "free will" constantly comes up in connection with the quantum enigma. So just for now, let's talk of this limited free will.

Free Will

Problems with free will arise in several contexts. Here's an old one: Since God is omnipotent, it might seem unfair that we be held responsible for *anything* we do. God, after all, had control. Medieval theologians resolved this issue by deciding that every train of events starts with a "remote efficient cause" and ends with a "final cause," both in God's hands. Causes in between come about through our free choices, for which we will be held accountable on judgment day.

This medieval concern is not completely remote from that of today's philosophers of morality. Similarly, criminal defense lawyers can make the concern practical by arguing that the defendant's actions were determined by genetics and environment rather than by free will. We, however, will deal with a more straightforward free-will issue.

Classical physics, Newtonian physics, is completely deterministic. An "all-seeing eye," viewing the situation of the universe at one time, can know its entire future. If classical physics applied to *everything*, there would be no place for free will.

However, free will can happily coexist with classical physics. In chapter 3 on the Newtonian worldview, we told how physics, in days gone by, could stop at the boundary of the human body, or certainly at the then completely mysterious brain. Scientists could dismiss free will as not their concern and leave it to the philosophers and theologians.

That dismissal does not come so easily today as scientists study the operation of the brain, its electrochemistry, and its response to stimuli. They deal with the brain as a physical object whose behavior is governed by physical laws. Free will does not fit readily into that picture. It lurks as a specter off in a corner.

Most neurophysiologists and psychologists tacitly ignore that corner. Some though, taking a physical model to apply broadly, deny that free will exists and claim that our *perception* of free will is an illusion. The controversy this creates will be right up front when we soon discuss the "hard problem" of consciousness.

How could you *demonstrate* the existence of free will? Perhaps all we have is our own feeling of free will and the claim of free will that others make. If no demonstration is at all possible, perhaps the existence of free will is meaningless. Here's a counter to that argument: Though you can't demonstrate your feeling of pain to someone else, you know it exists, and it's certainly not meaningless.

A famous free-will experiment has generated fierce argument. In the early 1980s, Benjamin Libet had his subjects flex their wrist at a time of their choice, but without forethought. He determined the order of three critical times: the time of the "readiness potential," a voltage that can be detected with electrodes on the scalp almost a second before any voluntary action actually occurs; the time of the wrist flexing; and the time the subjects reported that they had made their *decision* to flex (by watching a fast-moving clock).

One might expect the order to be (1) decision, (2) readiness potential, (3) action. In fact, the readiness potential *preceded* the reported decision time. Does this show that some deterministic function in the brain brought about the supposedly free decision? Some, not necessarily Libet, do argue this way. But the times involved are fractions of a second, and the meaning of the reported decision time is hard to evaluate. Moreover, since the wrist action is supposed to be initiated without any "preplanning," the experimental result seems, at best, ambiguous evidence against conscious free will.

In 2008, John-Dylan Haynes went beyond fractions of a second. He and his colleagues monitored neural activity with functional magnetic resonance imaging (fMRI). As letters appeared on a screen in front of them, subjects were asked to push the button in their right hand or the one in their left whenever they felt like it, or randomly. They then reported the letter they saw when they *decided* which button to push. From the fMRI signal, the researchers could predict seventy percent of the time (guessing works fifty percent of the time) the button that would be pushed, as much as ten seconds before the reported decision time. Haynes commented: "This doesn't rule out free will, but it does make it implausible."

Does it really? Presumably, if a subject were told *during* that ten-second interval, "You are going to push the left-hand button," they could still freely choose to push the right-hand button. Being able to roughly predict someone's behavior from an fMRI does not seriously challenge their free will. Predicting behavior from facial expression also works quite well.

Belief in our free will arises from our conscious perception that we make choices between possible alternatives. If free will is just an illusion, and we're all just sophisticated robots controlled by our neurochemistry with perhaps a bit of thermal randomness, is our consciousness then also an illusion? (If so, what is it that is *having* that illusion?)

Though it is hard to fit free will into our usual scientific worldview, we cannot, ourselves, with any seriousness, doubt it. J. A. Hobson's comment seems apt to us: "Those of us with common sense are amazed at the resistance put up by psychologists, physiologists, and philosophers to the obvious reality of free will."

If you're going to deny free will, stopping at the electrochemistry of the brain is arbitrary. After all, the motivation for suggesting such denial is the Newtonian determinism of classical physics. Being logically consistent, and thus accepting that reasoning all the way, we come to the *completely* deterministic world where the "all-seeing eye" can know the entire future of everything, including our experimenters' supposedly-free choice leading to the quantum enigma.

Unlike arbitrarily stopping at the electrochemistry of the brain, accepting *complete* determinism *does* evade the quantum enigma. For most of us, being "robots" in a completely deterministic world is too much to swallow. However, accepting both free will and the undisputed quantum experiments, we come to the quantum enigma. And to quantum theory for an explanation.

And quantum theory, unlike classical physics, is not a theory of the physical world independent of the experimenters' freely made decisions, their free will.

According to John Bell:

> It has turned out that quantum mechanics cannot be "completed" into a locally causal theory, at least as long as one allows . . . freely operating experimenters.

Before Bell's theorem, "free will"–or an explicit assumption of "freely operating experimenters"–was not something seen in a book about physics. It was certainly not seen in a serious physics *journal*. That's of course changing. In December 2010, for example, the prestigious journal *Physical Review Letters* published a calculation of precisely how much free will would have to be given up to account for the correlations observed by freely operating experimenters performing twin-state photon experiments. It's 14%. What that means in human terms is not clear.

Let's explore observation by Bell's "freely operating experimenters." Recall Pascual Jordan's defining statement of the Copenhagen interpretation, the working physicist's interpretation: "Observations not only *disturb* what is to be measured, they *produce* it." "Observation" here is an open-ended term, but the creation of physical reality by *any* kind of observation is hard to accept. However, it's not a new notion.

From Berkeley to Behaviorism

The idea of physical reality being created by its observation goes back thousands of years to Vedic philosophy, but we skip ahead to the eighteenth century. In the wake of Newton's mechanics, the materialist view that all that exists is matter governed by mechanical forces gained wide acceptance. Not everybody was happy with it.

The idealist philosopher George Berkeley saw Newtonian thinking as demeaning our status as freely choosing moral beings. Classical physics seemed to leave little room for God, and that appalled him. He was, after all, a bishop. (It was common in those days for English academics to be ordained as Anglican priests, though the celibacy of Newton's day was no longer required. Berkeley married.)

Berkeley rejected materialism with the motto *esse est percipi*, "to be is to be perceived," meaning all that exists is created by its observation. To the old question, "If a tree falls in the forest with no one around to hear it fall, is there any sound?" Berkeley's answer would presumably be that there wasn't even a tree were it not observed.

Though Berkeley's almost solipsistic stance may seem a bit batty, many idealist philosophers of his day were enthusiastic about it. Not so Samuel Johnson, who supposedly responded by kicking a stone, stubbing his toe, and declaring, "I refute him thus!" Stone kicking made little impression on those partial to Berkeley's thinking, which is, of course, impossible to disprove.

Though this is not quite Berkeley's position, here is a centuries-old limerick to illustrate the attention such ideas got:

> There was a young fellow named Todd
> Who said, "It's exceedingly odd
> To think that this tree
> Should continue to be
> When there's no one about in the Quad."

The reply:

> There is nothing especially odd;
> I am always about in the Quad.
> And that's why this tree
> Can continue to be
> When observed by
> Yours faithfully, God.

God may be omnipotent but, we note in the spirit of this limerick, He is not omniscient. If God's observation collapses the wavefunctions of large things to reality, quantum experiments indicate that He is not observing the small.

The idea that the world around us was being created by its observation never took hold. Most practical people, surely most scientists of the eighteenth century, considered the world to be made up of solid little particles, which some called "atoms." These were presumed to obey mechanical laws much as did those larger particles, the planets. While physical scientists might speculate about the mind, and some used hydraulic pictures for it instead of today's computer models, for the most part they ignored it.

In the nineteenth and much of the twentieth century, scientific thinking was generally equated with materialist thinking. Even in psychology departments, consciousness did not warrant serious study. Behaviorism became the dominant view. People were to be studied as "black boxes" that received stimuli as input and provided behaviors as output. Correlating the behaviors with the stimuli was all that science needed to say about what goes on inside. If you knew the behavior corresponding to every stimulus, you would know all there is to know about the mind.

The behaviorist approach had success in revealing how people respond and, in some sense, why they act as they do. But it did not even address the *internal* state, the feeling of conscious awareness and the making of apparently free choices. According to behaviorism's leading spokesman, B. F. Skinner, the assumption of a conscious free will was unscientific. But with the rise of humanistic psychology in the latter part of the twentieth century, behaviorist ideas seemed sterile.

The "Hard Problem" of Consciousness

Behaviorism had waned when, in the early 1990s, David Chalmers, a young Australian philosopher, shook up the study of consciousness by identifying the "hard problem" of consciousness. In a nutshell, the hard problem is that of explaining how the biological brain generates the subjective, inner world of *experience*. Chalmers's "easy problems" include such things as the reaction to stimuli and the reportability of mental states, and all the rest of consciousness studies. Chalmers does not imply that his easy problems are easy in any absolute sense. They are easy only relative to the hard problem. Our present interest in the hard problem of consciousness, or awareness, or experience, arises from its apparent similarity (and connection?) to the hard problem of quantum mechanics, the problem of observation.

Before going on about the hard problem and the heated arguments it continues to generate, a bit about David Chalmers: As an undergraduate student, he studied physics and mathematics and did graduate work in mathematics before switching to philosophy. Though it is not central to his argument, Chalmers considers quantum mechanics likely relevant to consciousness. The last chapter of his landmark book, *The Conscious Mind*, is titled "The Interpretation of Quantum Mechanics." David Chalmers was a faculty colleague at the University of California, Santa Cruz, in the philosophy department, before he (to our regret) moved to the University of Arizona to become a director of the Center for Consciousness Studies. He is, at the time of this writing, back in his native Australia as the director of the Centre for Consciousness at Australia National University.

Chalmers's easy problems often involve the correlation of neural activity with physical aspects of consciousness, the "neural correlates of consciousness." Brain-imaging technology today allows the detailed visualization of metabolic activity inside the thinking, feeling brain and has stimulated fascinating studies of thought processes.

Exploration of what goes on inside the brain is not new. Neurosurgeons have long correlated electrical activity and electrical stimulation with reports of conscious perception by placing electrodes directly on the exposed brain. This is done largely for therapeutic purposes, of course, and scientific experimentation is limited. Electroencephalography (EEG), the detection of electrical potentials on the scalp, is even older. EEG can rapidly detect neuronal activity but can't tell much about where in the brain the activity is taking place.

Positron emission tomography (PET) is better at finding out just where in the brain neurons are firing. Here, radioactive atoms, of oxygen for example, are injected into the blood stream. Radiation detectors and computer analysis can determine where there is an increase in metabolic activity, and can correlate this call for more oxygen with reports of conscious perceptions.

The most spectacular brain imaging technology is functional magnetic resonance imaging (fMRI). It is better than PET at localizing activity and involves no radiation. (The examined head must, however, be held still in a large, usually noisy, magnet.) MRI is the medical imaging technology we described in chapter 9 as one of the practical applications of quantum mechanics. fMRI can identify the part of the brain that is using more oxygen during a particular brain function responding to an external stimulus.

fMRI can correlate a brain region with the neural process involved in, say, memory, speech, vision, or reported awareness. The computer-generated, false-color brain images produced can display just which regions in a brain require more blood when someone thinks, say, of food or feels pain. Like any technique based on metabolic activity, fMRI is not fast.

Is the *physical* brain that these techniques observe, presumably all there *is* to the brain, also all there is to the mind? While the work today relating neural electrochemistry to consciousness may be rudimentary, just suppose that improved fMRI, or some future technology, could *completely* identify particular brain activations with certain conscious experiences. This would correlate all (reported) conscious feelings with metabolic activity, and perhaps even with the underlying electrochemical phenomena. Such a complete set of the neural correlates of consciousness is the ultimate goal of much of today's consciousness research involving the brain.

Were this goal actually achieved, some say we would have accomplished all that *can* be accomplished. Consciousness, they claim, would be completely explained because there is nothing to it *beyond* the neural activity we correlate with the experiences we *call* "consciousness." If we take apart an old pendulum clock and see how the swinging weight driven by a spring moves the gears, we can learn all there is to know about the workings of the clock. The claim here is that consciousness will be similarly explained by our learning all about the neurons making up the brain.

Francis Crick, physicist co-discoverer of the DNA double helix, who turned brain scientist, looked for the "awareness neuron." For him, our subjective experience, our consciousness, is nothing but the activity of such neurons. His book *The Astonishing Hypothesis* identifies that hypothesis:

> "You," your joys and sorrows, your memories and your ambitions, your sense of personal identity and free will, are in fact no more than the behavior of a vast assembly of nerve cells and their associated molecules.

If so, the intuition that our consciousness and free will are experiences beyond the mere functioning of electrons and molecules in our brain is an illusion. Consciousness should therefore ultimately have a reductionist explanation. It should, in principle at least, be completely describable in terms of simpler entities, the neural correlates of consciousness. Subjective feelings thus supposedly "emerge" from the electrochemistry of neurons. This is akin to the readily accepted idea that the surface tension or "wetness" of water emerges from the interaction of hydrogen and oxygen atoms forming contiguous molecules of H_2O.

Such emergence forms Crick's "astonishing hypothesis." Is it really so astonishing? We suspect that, to most physicists at least, it would seem a most natural guess.

Crick's long-time younger collaborator, Christof Koch, takes a more nuanced approach:

> Given the centrality of subjective feelings to everyday life, it would require extraordinary factual evidence before concluding that qualia and feelings are illusory. The provisional approach I take is to consider first-person experiences as brute facts of life and seek to explain them.

In a slightly different context, Koch further balances different views:

> While I cannot rule out that explaining consciousness may require fundamentally new laws, I currently see no pressing need for such a step.
>
> ... [But] [t]he characters of brain states and of phenomenal states [*experienced* states] appear too different to be completely reducible to each other. I suspect that their relationship is more complex than traditionally envisioned.

David Chalmers, a principal spokesperson for a point of view diametrically opposite to Crick's, sees explaining consciousness purely in terms of its neural correlates to be *impossible*. At best, Chalmers maintains, such theories tell us something about the *physical* role consciousness may play, but those physical theories don't tell us how consciousness arises:

> For any physical process we specify there will be an unanswered question: Why should this process give rise to [conscious] experience? Given any such process, it is conceptually coherent that it could ... [exist] in the absence of experience. It follows that no mere account of physical process will tell us why experience arises. The emergence of experience goes beyond what can be derived from physical theory.

While atomic theory might reductively explain the wetness of water and why it clings to your finger, that's a far cry from explaining your *feeling* of its wetness. Chalmers, denying the possibility of *any* reductive explanation of consciousness, suggests that a theory of consciousness should take experience as a primary entity alongside mass, charge, and space-time. He suggests that this new fundamental property would entail new fundamental laws, which he calls "psychophysical principles."

Chalmers goes on to speculate on these principles. The one he considers basic, and the one most interesting to us, leads to a "natural hypothesis: that information (at least some information) has two basic aspects, a physical aspect and a phenomenal aspect." This postulate of a dualism recalls the situation in quantum mechanics, where the wavefunction also has two aspects: On the one hand, it is the total physical reality of an object, while on the other hand, that reality, some have conjectured, is purely "information" (whatever that means).

To argue that conscious experience goes beyond intellectual knowing, some tell the story of Mary. Mary is a scientist of the future who knows *everything* there is to know about the perception of color. But Mary has never been outside a room where everything is black or white. One day she is shown something red. For the first time, Mary *experiences* red. Her experience of red is something *beyond* her complete knowledge of red. Or is it? You can no doubt generate for yourself the pro and con arguments the Mary story provokes.

Philosopher Daniel Dennett in his widely quoted book *Consciousness Explained*, describes the brain's dealing with information as a process where "multiple drafts" undergo constant editing, coalescing at times to produce experience. Dennett denies the existence of a "hard problem," considering it a form of mind–brain dualism. He claims to refute it by arguing:

> No physical energy or mass is associated with them [the signals from the mind to the brain]. How then do they make a difference to what happens in the brain cells they must affect, if the mind is to have any influence over the body? . . . This confrontation between quite standard physics and dualism . . . is widely regarded as the inescapable and fatal flaw of dualism.

Since Chalmers argues that consciousness obeys principles *beyond* standard physics, it is not clear that an argument *based* on "quite standard physics" can be a refutation of Chalmers. Moreover, there's a quantum loophole in Dennett's argument: No mass or energy is necessarily required to determine to *which* of the set of possible states a wavefunction will collapse upon observation.

Our own concern with the hard problem of consciousness arises, of course, because physics has encountered consciousness in the quantum enigma, which physicists call the "measurement problem." Here, aspects of physical observation come close to those of conscious experience. In both cases, something *beyond* the normal treatment, of physics, or of psychology, appears to be needed for a solution.

The essential nature of the measurement problem in quantum mechanics has been in dispute since the inception of the quantum theory. Similarly, ever since consciousness has become scientifically discussed in psychology and philosophy, its essential nature has been in dispute. An example of the rather extreme divergence of opinion appeared in 2005 in the *New York Times*, where some leading scientists were asked to state their beliefs. According to cognitive scientist Donald Hoffman:

> I believe that consciousness and its contents are all that exists. Space-time, matter and fields never were the fundamental denizens of the universe but have always been, from their beginning, among the humbler contents of consciousness, dependent on it for their very being.

Psychologist Nicholas Humphrey sees it differently:

> I believe that human consciousness is a conjuring trick, designed to fool us into thinking we are in the presence of an inexplicable mystery.

One way to explore the nature of consciousness, and its existence, is to ask who or what can possess it.

A Conscious Computer?

We each know *we* are conscious. Perhaps the only evidence for believing that others are conscious is that they more or less look like us and behave like us. What other evidence is there? The assumption that our fellow humans are conscious is so ingrained that it is hard to express the reasons for our believing it.

How far down does consciousness extend? What about cats and dogs? What about earthworms or bacteria? Some philosophers see a continuum and even attribute a bit of consciousness to a thermostat. On the other hand, maybe consciousness turns on abruptly at some point on this scale.

After all, Nature can be discontinuous: going below 32°F, liquid water abruptly becomes solid ice.

Let's step back from consciousness and just talk about "thinking," or intelligence. Today, computer systems called artificial intelligence, or AI, assist doctors in diagnosing disease, generals in planning battles, and engineers in designing yet better computers. In 1997, IBM's Deep Blue beat the world chess champion, Garry Kasparov.

Did Deep Blue *think*? It depends on what you mean by thinking. Information theorist Claude Shannon, when asked whether computers will ever think, supposedly replied: "Of course. I'm a computer, and I think." But the IBM scientists who designed Deep Blue insist that their machine is just a fast calculator evaluating a hundred million chess positions in the blink of an eye. Whether or not Deep Blue thinks, it is surely not conscious.

But if a computer *appeared* conscious in every respect, wouldn't we have to accept it as conscious? We follow the time-honored principle that if it looks like a duck, walks like a duck, and quacks like a duck, it must be a duck.

The interesting question is whether it is possible to *build* a conscious computer, and therefore a conscious robot. Computer consciousness is sometimes called strong artificial intelligence, or "strong AI." (Would it be murder to pull the plug on a truly conscious robot?) Logical "proofs" have been advanced that strong AI is, in principle, possible. There are other "proofs" that it is impossible. How could you tell if a computer were conscious?

In 1950 Alan Turing proposed a test for computer consciousness. He actually called it a test for whether a computer could think; a scientist wouldn't use the term "consciousness" back then. (Turing also designed the first programmed computer and developed a theorem for what computers could ultimately do, or not do. Turing was later arrested for his homosexuality, and in 1954 committed suicide. Many years after his death, officials revealed that it was Turing who broke Germany's Enigma code. The Allies were thus able to read the enemy's most secret messages, probably shortening World War II by many months.)

The Turing test uses essentially the same criterion for deciding whether a computer is conscious as we do in ascribing consciousness to another individual: Does it look and behave more or less like me? Let's not worry about the "look" part; a human-*looking* robot can no doubt be accomplished. The issue is whether its computer brain gives it consciousness.

To test whether a particular computer is conscious, it should, according to Turing, be enough to communicate with it by a keyboard and carry on any conversation, for as long as you wish. If you can't tell whether you are communicating with a computer or another human, it passes the Turing test. Some would then say that you cannot deny that it is conscious.

In class one day, I (Bruce) casually commented that any human could easily pass a Turing test. One young woman objected: "I've *dated* guys who couldn't pass a Turing test!"

Consciousness is a mystery we explore because physics' encounter with it presents us with the quantum enigma. In our next chapter, the mystery meets the enigma.

17

The Mystery Meets the Enigma

When the province of physical theory was
extended to encompass microscopic phenomena
through the creation of quantum mechanics, the
concept of consciousness came to the fore again:
It was not possible to formulate the laws of
quantum mechanics in a fully consistent way
without reference to the consciousness.

—Eugene Wigner

When there are two mysteries, it is tempting to
suppose that they have a common source. This
temptation is magnified by the fact that the
problems in quantum mechanics seem to be deeply
tied to the notion of observership, crucially
involving the relation between a subject's
experience and the rest of the world.

—David Chalmers

Consciousness and the quantum enigma are not just two mysteries; they are *the* two mysteries: The first, the experimental demonstration of the quantum enigma, presents us with the mystery of the objective, physical world "out there," and the second, conscious awareness, presents us with the mystery of the subjective, mental world "in here." Quantum mechanics appears to connect the two.

The Encounter "Officially" Proclaimed

In his 1932 treatment, *The Mathematical Foundations of Quantum Mechanics*, John von Neumann rigorously displayed quantum theory's inevitable encounter with consciousness. Von Neumann considered an idealized quantum measurement starting with a microscopic object in a superposition state and ending with the observer. A Geiger counter, for example, completely isolated from the rest of the world, contacts a quantum system, say, an atom simultaneously in two boxes. The Geiger counter is set to fire if the atom is in the bottom box, and to remain unfired if the atom is in the top box. Von Neumann showed that the isolated Geiger counter, a physical object governed by quantum mechanics, would *entangle* with the atom in both boxes. It would thus be in a superposition state with the atom. It would thus be simultaneously in the fired and unfired state. (We saw this situation in the case of Schrödinger's cat.)

Should a second device, also isolated, contact the Geiger counter—say an electronic instrument indicating whether or not the Geiger counter has fired—it joins the superposition state wavefunction, indicating both situations simultaneously. This so-called "von Neumann chain" can continue indefinitely. Von Neumann showed that *no* physical system obeying the laws of physics (i.e., quantum theory) could collapse a superposition state wavefunction to yield a particular result. However, we know that the observer at the end point of the von Neumann chain always sees a *particular* result, a fired or not fired Geiger counter, not a superposition. Von Neumann showed that for all practical purposes the wavefunction could be *considered* collapsed at any macroscopic stage of the measurement chain where an interference demonstration becomes essentially impossible. Nevertheless, he concluded that, strictly speaking, collapse takes place only at the "Ich," the same word Freud used for the Ego, the conscious mind.

A couple of years later Schrödinger told his cat story to illustrate the "absurdity" of his own quantum theory. His story was essentially based on von Neumann's conclusion that in principle quantum theory requires a conscious observation, consciousness, in order to collapse a superposition state. Could this really be so?

Do We *Need* a Conscious Observer?

Is a conscious observer *needed* to collapse a wavefunction? One can defend either a "yes" or a "no" answer to this question. Both "collapse" and "consciousness" admit a broad spectrum of meanings. For Schrödinger's cat story, we can, with the Copenhagen interpretation, consider the wavefunction of the macroscopic Geiger counter to have collapsed to the fired or not-fired state as soon as the atom encountered it. The cat would then quickly become dead or alive, never entering a superposition state. On the other hand, since the Geiger counter and the rest of Schrödinger's "hellish contraption" were isolated from the environment, perhaps the simultaneous living and dead cat did *not* become living or dead until an observer became conscious of its state as *either* living or dead.

That latter case gets complicated. "Becoming conscious" can mean seeing the cat and achieving a full awareness that Schrödinger's cat is, say, dead. On the other hand, "conscious observation" of the cat's state could consist of seeing a flash of light come through holes in the box, a flash that would not come through were the cat standing. (Is the *cat* a conscious observer? For this argument's sake, consider a robot-cat, which falls if the Geiger counter fires.)

An observer who knew the significance of the flash would have consciously observed the cat to be dead. However, what if the observer was merely conscious of a flash of light with it having no particular meaning? Or what if the observer, hit by the flash, was totally unaware of it? That observer would nevertheless be entangled with those photons and thus entangled with the cat. If *that* entanglement with a conscious observer constitutes collapse, we have greatly broadened the meaning of "consciousness."

We emphasize that this question of the collapse of the wavefunction arises out of the quantum *theory*. "Collapse" and "wavefunction" are terms of the theory. The quantum enigma arises directly from quantum *experiment,* through the free choice of the experimenter. The enigma arises without any need to talk of "collapse" or "wavefunction."

Conscious Awareness versus Entanglement

Consider once again an atom in a box pair. Let the boxes be transparent to light. Were the atom in the top box, a photon sent through the top box would bounce off the atom in a new direction. Were the atom in the bottom box, the photon would come straight through. With the atom in a superposition state in both boxes, the photon entangles with the atom, joining the superposition state. It both bounces off and comes straight through. An *isolated* Geiger counter on that straight-through path would entangle with the photon-atom wavefunction and be both fired and unfired.

Now, however, suppose the Geiger counter, hit by the photon, sits on a table that rests on the floor. This *not*-isolated counter interacts with the table (as its atoms bounce against the table's atoms). It is thus *entangled* with the table, and therefore with the rest of the world, which includes people. The atom, entangled with the photon, entangled with the counter, is now entangled with conscious observers. If no one looks at the Geiger counter (or knows what its firing means), no one *knows* which box the atom is in.

Does this entanglement of the atom with the rest of the world, including conscious observers, collapse the atom wholly into a single box? Or does the collapse into a single box require conscious *awareness* of which box the atom is in by an actual look at the Geiger counter? How could we possibly tell? Strictly speaking, unless we invoke something beyond our present quantum theory, the atom is perhaps still in both boxes, and the Geiger counter is both fired and not fired. (In chapter 13 we spoke of the proposed experiment with observers in space to test this.)

The rest of the world instantaneously *entangles* with our photon as soon as it hits the *not*-isolated Geiger counter. According to quantum theory, entanglement travels infinitely fast. But for a remote person to become *aware* of the condition of the counter, he or she would have to communicate by some *physical* means, which could not exceed the speed of light.

We saw entanglement traveling faster than light, infinitely fast, presumably, in Bell's theorem experiments. *Immediately* upon the observation of the polarization of one twin-state photon, the polarization of its twin

Figure 17.1 Drawing by Nick Kim, 2000. © American Institute of Physics

is set. That's entanglement. But only when the two observers gain awareness of each other's result can they know whether they had a match or mismatch. A photon "learns" its twin's behavior instantaneously, but Alice and Bob can become aware of each other's result only at a rate limited by the speed of light.

Figure 17.1 is a cartoon strip from *Physics Today* in May 2000 that's relevant in a few ways. (When the quantum enigma comes up in physics journals, other than to supposedly resolve the issue, it's often treated with humor.) Chris, being entangled with Eric and the rest of the world, would, of course, not go into a "superposition of all possible states" when Eric looks away. After all, an atom you found in a particular box would not go into a superposition in both boxes when you looked away.

Consciousness and Reduction

With consciousness encountered in the quantum experiment, or even just arising in the quantum theory, we can see a problem with reductionism. The reductionist perspective seeks to reduce the explanation of a complex system to its underlying science. For example, one can seek explanations of psychological phenomena in biological terms. Biological phenomena can be seen as ultimately chemical. And no chemist doubts that chemical phenomena are fundamentally the interactions of atoms obeying quantum physics. Physics, itself, can supposedly rest firmly on primitive empirical ground.

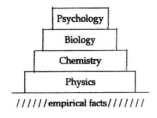

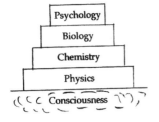

Figure 17.2 Hierarchy of scientific explanation revisited

In chapter 3 we represented this view with the reductionist pyramid. That view of the primitive empirical ground on which physics rests is challenged by quantum mechanics, where physics ultimately rests on observation. Observation somehow involves consciousness, *whatever* that is. Therefore, add a somewhat cloudy consciousness at the base of our reductionist pyramid in Figure 17.2. For all practical purposes, science will always be hierarchical, with each level in the hierarchy needing its own set of concepts. Nevertheless, this new perspective on reduction can change the way we perceive the scientific enterprise.

The Robot Argument

A mechanistic robot argument is often presented to deny an encounter with consciousness. Here's how that argument goes: We don't need a *conscious* observer to collapse a wavefunction, because a *not*-conscious robot can do the same thing. The robot is presented with sets of box pairs prepared as described in chapter 7. It is programmed to do either a "which-box" experiment or an "interference" experiment with each set and print out a report of its results. The printout would be indistinguishable from a printout presented by a conscious experimenter. Since no consciousness was involved in the experiments, no enigma involving consciousness exists.

Here's why this argument does not work. The robot's printout indicates that with certain sets it did a "which-box" experiment establishing that *these* sets contained objects wholly in a single box. With other sets, it did an "interference" experiment establishing that *those* sets contained objects distributed over both boxes. The robot's printout tells that the different box-pair sets indeed *contained* objects of just such different kinds.

A problem: How did the robot "decide" to do the *appropriate* experiment with each box-pair set? If it did an "interference" experiment with objects actually wholly in a single box, it would have gotten no pattern, just a uniform distribution. That *never* happened. And what would it report if it did a "which-box" experiment with objects distributed over both boxes? A partial object is *never* reported.

You investigate the robot's inexplicably always-appropriate choice of experiment. You find it used a choice method as effective as any available to a mechanical robot, a coin flip. Heads, it did the "which-box" experiment; tails, the "interference" experiment. The robot's appropriate choices arose from the coin's landing being connected with what was presumably actually in a particular box-pair set. You find this connection inexplicable, mysterious.

Continuing your investigation, you replace the robot's coin flipping by the decision method you are *sure* is not correlated with what was actually in a particular box-pair set: *your own consciously made free choice.* You now push a button telling the robot which experiment to do with each box-pair set. You find that by your free choice you can establish *either* that the objects were concentrated in a single box *or* that they were distributed over both boxes, *either* of two contradictory things. You are now faced with the quantum enigma, and consciousness is encountered. The robot argument denying the encounter with consciousness does not work.

Refutation of the robot argument indeed requires accepting our conscious perception that we *can* make free choices, that our choices *can*, in part at least, be independent of what exists in the external physical world. The alternative is that we are robots in a totally deterministic world.

The Only *Objective* Evidence for Consciousness

By "objective evidence" we mean third-person evidence that can be displayed to essentially anyone. Objective evidence in this sense is the normal requirement for establishing a scientific theory. We each *know* we are conscious; that's first-person evidence for consciousness. Others report they are conscious; that's second-person evidence. Without third-person evidence, objective evidence that consciousness itself can directly involve something physically observable, its very existence is deniable. It *is* sometimes denied.

Some claim that "consciousness" is no more than a name for the electrochemical behavior of the vast assembly of nerve cells and their associated molecules in our brains. Can some direct role for consciousness be displayed beyond the electrochemical aspects confined within our bodies?

What might qualify as objective evidence for consciousness directly involving the physical? The two-slit experiment, or its box-pair version, might seem to *almost* qualify. One weakness is that the evidence is *circumstantial* evidence rather than *direct* evidence. That is, one fact (the interference pattern depends on box-pair spacing) is used to establish a second fact (the object had been in both boxes).

Circumstantial evidence can be convincing beyond a reasonable doubt. It can, for example, legally secure a conviction. But the logic of circumstantial evidence can be circuitous. Therefore, in the spirit of our Neg Ahne Poc story, we first present an example, one that does *not* really exist, but one that would display *direct* evidence for consciousness. Its *direct* evidence is easy to analyze, and its *analogy* with our quantum experiment puts that *circumstantial* evidence more squarely in front of us. Here's our story:

You are presented with a set of box pairs. Choosing to open the boxes of a pair one at a time, you invariably find that one box contains a marble and the other is empty. The marble is randomly in the first box you open or the second, and the other box of the pair is empty. On the other hand, if you choose to open the boxes of each pair at about the same time, you always find half a marble in each box.

You, and the team of experts you hire with your unlimited budget, search for any evidence that the physical process of box-pair opening could possibly have any effect on the condition of the marble. You establish beyond a reasonable doubt that no such physical effect exists.

Of course, this demonstration *cannot* be done. But *if* it could, you would have little alternative but to accept it as objective evidence that the *conscious choice* of opening technique can affect a physical situation. This would be *objective*, third-person evidence (though not proof) that consciousness exists as an entity beyond its neural correlates.

The archetypal quantum experiment, the two-slit experiment or our box-pairs experiment, comes close to this demonstration. No physical effect of box-pair opening can be found. Your conscious choice of which experiment to do ("which-box", or "interference") can apparently create *either* of two contradictory physical situations in the box pairs. Since the demonstration is displayable to anyone, the quantum experiment is *objective* evidence.

Though the quantum experiment must involve interference, and is therefore at best *circumstantial* evidence, it is the only *objective* evidence for consciousness we have. Evidence, of course, is not proof. The quantum experiment is the suspicious footprint at the crime scene *suggesting* a culprit.

Does the quantum experiment actually show consciousness reaching out and doing something physical? In serious moments as physicists, we can't even half-believe that. But developer of quantum theory and Nobel laureate Eugene Wigner has speculated:

> Support [for] the existence of an influence of the consciousness on the physical world is based on the observation that we do not know of any phenomenon in which one subject is influenced by another without exerting an influence thereupon. This appears convincing to this writer. It is true that under the usual conditions of experimental physics or biology, the influence of any consciousness is certainly very small. "We do not need the assumption that there is any such effect." It is good to recall, however, that the same may be said of the relation of light to mechanical objects. . . . It is unlikely that the [small] effect would have been detected had theoretical considerations not suggested its existence. . . .

This kind of speculation can enrage some physicists. But you at least know the undisputed experimental facts upon which Wigner's wild speculation is based.

Position Is Special

Why can't we *see* an object simultaneously in two boxes? Quantum theory provides no answer. Strictly speaking, an object wholly in Box A can *also* be considered to be in a "superposition state." It is in a superposition (or sum) of the state {in Box A + in Box B} plus the state {in Box A − in Box B}. These just add up to {in Box A}. Similarly, the living-cat state is a superposition of the state {living + dead} plus the state {living − dead}. The missing factor of 2 is accounted for in the actual mathematics of quantum theory.

All these states have equivalent status as far as quantum theory is concerned. Why, then, do we always see things in certain kinds of states, states characteristic of a particular position? We never actually see the weird states corresponding to things being simultaneously in different positions. (Schrödinger's simultaneously living-and-dead cat is such a weird state because to distinguish the living state from the dead state, some atoms in a living cat must be in different positions than the atoms in a dead cat.)

For our object in a box pair, we inferred that it had been simultaneously in two boxes by doing an interference experiment. But our actual *experiences* in the interference experiment were the *positions* of objects in particular maxima of an interference pattern.

Arguably, the reason we observe only states characterized by unique positions is that we humans are beings who can experience *only* position (and time). Speed, for example, is position at two different times. When we see things with our eyes, it is because of light on particular positions on our retina. We feel by touch the position of something on our skin; we hear by the changing position of our eardrums; we smell by the effects on certain receptor positions in our nose. We therefore build our measuring instruments to display their results in terms of position—typically, the position of a meter pointer or of a light pattern on a screen. Nothing in quantum theory forces this situation. We humans seem constructed in this special way.

Is it conceivable that other beings could experience reality differently? Could they possibly directly experience the superposition states whose existence we only infer? To them, an atom simultaneously in both boxes, or Schrödinger's cat simultaneously alive and dead, would be "natural." That is, after all, the quantum way, presumably Nature's way. They would therefore experience no measurement problem, no quantum enigma.

Two Enigmas

We can actually see two measurement problems, two enigmas. We focused on observer-created reality: observation causing, say, the looked-at atom to appear wholly in a single box, or Schrödinger's cat to be either alive or dead. (Einstein's quip that he believed the moon was really there even when no one was looking challenged this enigma.) A less disturbing enigma is Nature's randomness: How does it come about that the atom *randomly* appears in, say, Box A rather than Box B? How does the cat *randomly* come to be in, say, the alive state? (Einstein's quip that God doesn't play dice challenged this enigma.)

With Everett's many-worlds interpretation of quantum mechanics, you choose all possible experiments and see all possible results. According to this view, the "you" in one *particular* world, is troubled by the two enigmas only because you do not realize that at every observation, at every decision, you split and simultaneously exist in a multitude of different worlds. From an Everettian point of view, the *complete* "you" should experience neither enigma.

Let's contrast the two enigmas with a bit of fantasy (inspired by a parable by Roland Omnès). On the higher plane on which they dwell, Everettians happily experience the multitude of simultaneous realities given by quantum theory. No enigmas trouble them. One young Everettian, sent down to explore planet Earth, was shocked to find his simultaneous multiple realities collapse to a single actuality (like a wavefunction collapsing to be wholly in a single box). His curiosity impelled repeated descents. Each time he saw his realities randomly collapse to one of the many he was accustomed to perceive simultaneously on his higher plane. Baffled by this collapse, something not explicable within the quantum theory he understood so well, he reported an enigma: Down on Earth, Nature randomly selects a single actuality.

Our Everettian had a favorite way of looking at the multiple realities he could experience (like our choice of which experiment to do with the box pairs). He understood, however, that this personal choice, what physicists call a "basis," was, according to quantum theory, equivalent to any other. In a rather unusual mood on a particular descent to Earth, our

Everettian adopted a different basis for his multiple realities. He experienced a second bafflement. The random collapse was not merely to a specific actuality, something he had by now gotten accustomed to, but to an actuality that was logically inconsistent with one presented by his previous way of looking. He had to report a second, and more troubling, enigma: Down on Earth, his conscious choice of the way he looked could create *inconsistent* realities.

Two Quantum Theories of Consciousness

Theories encompassing both mind and matter that go beyond analogy must be big and bold. They are inevitably controversial. The Penrose-Hameroff approach is based on quantum gravity, which is a still-being-developed theory describing black holes and the Big Bang, to which Roger Penrose is a major contributor. The Penrose-Hameroff approach to consciousness also involves ideas from mathematical logic and neuronal biology.

The mathematician Kurt Gödel proved that any logical system contains propositions whose truth cannot be proven. We can, however, by insight and intuition, know the answer. Penrose controversially deduces from this that conscious processes are non-computable. That is, no computer can duplicate them. Penrose thus denies the possibility of strong artificial intelligence, or strong AI. If there can be no strong AI, consciousness, like the quantum enigma, goes beyond anything our *present* science can explain.

Penrose proposes a physical process beyond present quantum theory that rapidly collapses macroscopic superpositions to actualities. It causes a macroscopic object simultaneously in both Box A *and* Box B to rapidly become either in Box A *or* Box B. It causes Schrödinger's cat, simultaneously both alive *and* dead, to rapidly become either alive *or* dead. In general, it causes "and" to become "or." This process collapses, or "reduces," the wavefunction objectively, that is, for everybody, even without an observer. Penrose calls this process "objective reduction," abbreviated OR. He notes the appropriateness of the OR acronym. It brings about the "or" situation.

Penrose speculates that OR occurs spontaneously whenever two space-time geometries, and therefore gravitational effects, differ significantly. Stuart Hameroff, an anesthesiologist, who points out that he regularly turns consciousness off and then back on, suggested how this process might occur in the brain. Two states of certain proteins (tubulins) that exist within neurons might display Penrose's OR on a time scale appropriate for neural functions. Penrose and Hameroff claim that superposition states and long-range quantum coherence might exist within a brain, even though it is in physical contact with the environment, and that spontaneous ORs could regulate neural functions.

Such objective reductions, ORs, would constitute "occasions of experience." If entangled with objects external to the observer, an OR in the brain would collapse the wavefunction of observed objects, and everything entangled with them.

The three bases of the Penrose-Hameroff theory—noncomputability, the involvement of quantum gravity, and the role of tubulins—are each controversial. And the entire theory has been derided as having the explanatory power of "pixie dust in the synapses." However, unlike almost all other theories of consciousness, quantum or otherwise, it proposes a specific physical mechanism, some fundamental aspects of which are testable with today's technology. Such tests are under way, though the results are in dispute.

With another theory, Henry Stapp argues that *classical* physics can *never* explain how consciousness can have any physical effect, but that an explanation comes about naturally with quantum mechanics. We saw earlier how free will was permitted in deterministic classical physics only by excluding the mind from the realm of physics. Stapp notes that extending classical physics to the brain/mind would have our thoughts controlled "bottom-up" by the deterministic motion of particles and fields. Classical physics allows no mechanism for a "top-down" conscious influence.

Stapp takes off from von Neumann's formulation of the Copenhagen interpretation. Von Neumann, recall, showed that in viewing a microscopic object in a superposition state, the entire measurement chain, from, say, the atom to the Geiger counter, to the human eye looking at it, to the thus-entangled synapses in the observer's brain, must, strictly speaking, be considered part of a grand superposition state. Only consciousness, something beyond the Schrödinger equation and beyond present physics, can, according to von Neumann, collapse a wavefunction.

Stapp postulates two realities, a physical and a mental. The physical includes the brain, perhaps in a particular quantum superposition state. The mental includes one's consciousness and, in particular, one's intentions. That mental reality can intentionally act on the physical brain to choose a *particular* superposition state, which then collapses to one actual situation. Consciousness does not directly "reach out" to the external world in this theory, but this mental choice nevertheless determines, in part, the character of the physical world external to the body. It determines, for example, whether an object was wholly in a single box of its pair or simultaneously in both. The final random aspect of the choice (in which particular box, or in which maximum of the interference pattern the object is found, for example) is then made by Nature.

How can a large, warm brain remain in a particular quantum state long enough for a person's intentions to influence it? Random thermal motions of atoms in the brain would be expected to allow a quantum state to exist for only a very much shorter time than needed for mental processing. Stapp answers this with the demonstrated "quantum Zeno effect" (named for a Zeno-like claim: A watched pot never boils). When an unobserved atom, or any quantum system, decays from an upper state to a lower, the decay starts very slowly. If the system is observed very soon after the decay has started, it will almost certainly be found in the original state. The decay then starts over again from the original state. If the system is observed almost constantly, it almost never decays. Stapp applies this to one's mental intentions "observing" one's brain and thus holding it in a given quantum state for a sufficient time.

Stapp cites various psychological findings as evidence for his theory. The theory is, of course, controversial.

The Psychological Interpretation of Quantum Mechanics

Though quantum theory is outrageously counterintuitive, it works perfectly. Since Nature need not behave in accord with our intuition, is the measurement problem, the quantum enigma, just in our heads? Maybe so. But, if so, why do we find quantum mechanics so hard to accept? Why do the observed facts produce such a strong cognitive dissonance pitting our sense of free will against our belief in a physically real world existing independently of its observation?

Merely to say that we evolved in a world where classical physics is a good approximation is not enough. We evolved in a world where the sun apparently moved across the sky while the Earth stood still. Nevertheless, the once-counterintuitive Copernican picture is readily accepted despite our evolution. We also evolved in a world where things move slowly compared to the speed of light. Einstein's relativity can be grossly counterintuitive. Though it is difficult for physics students to initially accept that time passes more slowly in a moving rocketship, they soon adjust their intuition to do so. There are no "interpretations" of relativity. The more deeply you think about relativity, the *less* strange it seems. The more deeply you think about quantum mechanics, the *more* strange it seems.

What is it about the organization of our brain that makes quantum mechanics seem so weird? With this question, most physicists might assign the quantum enigma to psychology. Our unease with physical reality being created by its observation is then merely a psychological hang-up. That would be the psychological interpretation of quantum mechanics. The quantum enigma is then no longer a problem for physics. It's psychology. Perhaps it is something psychologists might actually address.

Does Quantum Mechanics Support Mysticism?

It is sometimes implied that the sages of ancient religions intuited aspects of contemporary physics. The argument can go on to claim that quantum mechanics provides evidence for the validity of these mystic teachings. Such reasoning is not compelling.

However, while the Newtonian worldview is sometimes seen as denying the *possibility* of any such ideas, quantum mechanics, telling of a universal connectedness and involving observation in the nature of reality, denies that denial. In this most general sense, one can see the findings of physics supporting certain thinking of ancient sages. (When Bohr was knighted, he put the Yin-Yang symbol in his coat of arms.)

Quantum mechanics tells us strange things about our world, things that we do not fully comprehend. The strangeness has implications beyond what is generally considered physics. Physicists might therefore be tolerant when non-physicists incorporate quantum ideas into their own thinking.

We physicists are, however, disturbed, and sometimes embarrassed, by the misuse of quantum ideas, as, for example, a basis for certain medical or psychological therapies (or investing schemes!). A touchstone test for misuse is the presentation of these ideas implying that they are *derived* from quantum physics rather than merely analogies suggested by it.

Quantum mechanics can, however, provide good jumping-off points for imaginative stories. The teleportation in *Star Trek* ("Beam me up, Scotty.") is an imaginative but acceptable extrapolation of the transmission of quantum influences in EPR-type experiments. Such stories are fine, if it is clear, as it is in *Star Trek*, that they are fiction. Unfortunately, that is not always so.

Analogies

Whether or not consciousness can have direct impact beyond the brain, quantum physics provides some compelling analogies. Though analogies, of course, *prove* nothing, they can stimulate and guide thinking. Analogies with Newton's mechanics sparked the Enlightenment. Here's a very general one by Niels Bohr:

> [T]he apparent contrast between the continuous onward flow of associative thinking and the preservation of the unity of the personality exhibits a suggestive analogy with the relation between the wave description of the motions of material particles, governed by the superposition principle, and their indestructible individuality.

Here are a few more that others have suggested:

Duality: It is often argued that the existence of conscious experience cannot be deduced from the physical properties of the material brain. Two qualitatively different processes seem to be involved. Similarly, in quantum theory, an actual event comes about not by the evolving wavefunction, but by the collapse of the wavefunction by observation. Two qualitatively different processes seem to be involved.

"Nonphysical" influences: If there's a "mind" that's other than the physical brain, how does it communicate with the brain? This mystery recalls the connection of two quantum-entangled objects with each other, by what Einstein called "spooky actions" and Bohr called "influences."

Observer-created reality: Berkeley's "to be is to be perceived" is the preposterous solipsistic view of all reality being created by observation. It is, however, reminiscent of what happens with our object in a box pair, or with Schrödinger's cat.

Observing thoughts: If you think about the content of a thought (its position), you inevitably change where it is going (its motion). On the other hand, if you think about where it is going, you lose the sharpness of its content. Analogously, the uncertainty principle shows that if you observe the position of an object, you disturb its motion. On the other hand, if you observe its motion, you lose the sharpness of its position.

Parallel processing: Neuronal action rates are billions of times slower than those of computers. Nevertheless, with complex problems, human brains can out-perform the best computers. The brain presumably achieves its power by working on many paths simultaneously. It's just such massively parallel processing that computer scientists attempt to achieve with quantum computers, whose elements are simultaneously in superpositions of many states.

The analogies between consciousness and quantum mechanics lead one to expect that an advance in the fundamentals of one field will stimulate an advance in the other. Analogies might even suggest testable connections of the two.

Paraphenomena

Paraphenomena are presumed happenings that are inexplicable within normal science. Three examples involve the mind: extrasensory perception (ESP), the acquisition of information by some means other than the normal senses; precognition, discerning what will happen in the future; and psychokinesis, the causing of a physical effect by mental action alone.

According to polls, most Americans (and English) have significant beliefs in the reality of such phenomena. When asked with a positive spin, "Do you think it likely that at least a *little* ESP exists?" more than half the students in a large general physics class raised their hands. (The two of us would answer "*not* likely.")

Since paraphenomena are often linked with the mysteries of quantum mechanics, they call for comment here. The link can be misleading, and is sometimes fraudulent. Such connection can embarrass physicists, as we can personally attest. It's one reason discussion of the quantum enigma is avoided by physicists.

There are, however, competent researchers claiming to display such phenomena. Out-of-hand dismissal, though common, can display preconception, and seem arrogant. It's demonstrably ineffective.

We cite one recent example of a report of paraphenomena to be taken seriously: in January 2011, the *New York Times* published an article titled "Journal's Paper on ESP Expected to Prompt Outrage." It did.

The paper, accepted for publication by one of the most respected psychology journals, is by Daryl Bem, a distinguished psychologist and professor at Cornell University. Bem reports extensive experimental evidence for ESP and precognition. Recognizing that paraphenomena violate the normal scientific worldview, Bem reminds his readers that: "Several features of [undisputed] quantum phenomena are themselves incompatible with our everyday conception of physical reality."

Scientists are supposed to be open-minded to what they see, even open to the hard-to-believe. Some scientists, *too* open to what they see, have deceived themselves with paraphenomena experiments. Magicians, on the other hand, being experts in deception, are not easily deceived. Magicians have famously exposed the flaws of some scientists who claimed evidence of paraphenomena. We note that psychologist Bem is an accomplished magician, and therefore less likely to be so deceived.

Hard-to-believe things require strong evidence. As yet, evidence for the existence of paraphenomena strong enough to convince skeptics does not exist.

But if—*if!*—any such phenomenon were convincingly demonstrated, demonstrated to initially skeptical scientists (and magicians), we would know where to start looking for an explanation: Einstein's "spooky actions." Going a bit further, the *demonstrated* existence of quantum phenomena expands the scale of what is conceivable, and thus increases the subjective likelihood of paraphenomena. ("Subjective" in the Bayesian probability sense.) The extreme unlikelihood of paraphenomena within present physical theory means that any confirmation, *no matter how weak an effect*, would force a radical change in our worldview.

In the following chapter we consider the implications of the quantum enigma on the grandest scale of all, the entire universe.

18

Consciousness and the Quantum Cosmos

In the beginning there were only probabilities.
The universe could only come into existence if
someone observed it. It does not matter that the
observers turned up several billion years later.
The universe exists because we are aware of it.

—Martin Rees

Martin Rees, Cambridge University professor and England's Astronomer Royal, surely did not mean the above quote to be taken literally. Having come this far in the book, you at least know what might stimulate such a comment. It's a big step from the small things for which observer-created reality has been demonstrated to the whole universe. Quantum theory, however, supposedly applies to everything.

Quantum theory presumably encompasses most of the phenomena of physics (and of biology). It is only mysteries posed by the quantum experiment, and mysteries posed by cosmology, that seem to require totally new concepts. We've seen leading quantum cosmologists, Wigner, Penrose, and Linde, each suggesting that consciousness will, in some sense, be encountered in seeking those new concepts. Feeling the same way, we can't leave our discussion without a chapter on the cosmos.

Einstein's theory of gravity, "general relativity," appears to work perfectly for the large-scale universe. It also predicts black holes, and is required for dealing with the Big Bang. However, understanding black holes and the Big Bang also requires dealing with things at the small scale. It therefore requires quantum theory. Requiring *both* general relativity *and* quantum theory poses a problem: General relativity and quantum theory resist connection.

The problem is that quantum theory assumes a fixed stage of space and time, or space-time, and then describes the motion of matter on this stage. But in general relativity, the stage warps as matter tells space how to curve, and space tells matter how to move. String theorists and others have struggled for decades, and still vainly struggle, to couple these two fundamental descriptions of Nature to produce a quantum theory of gravity.

When, several years ago, I told a string-theorist colleague of my interest in the quantum enigma, "Bruce, we're not ready for that," was his response. His point was that a resolution of what he would call the quantum measurement problem, likely required still-to-come advances in quantum gravity theory—and, he felt, they would in no way involve consciousness. Perhaps. But today's cosmology, our view of the universe as a whole, presents a quantum enigma, one seeming to involve consciousness, on an ever-grander scale.

Black Holes, Dark Energy, and the Big Bang

Black Holes

When a star exhausts the nuclear fuel that keeps it hot and therefore expanded, the star collapses under its gravitational self-attraction. If its mass exceeds a certain critical amount, no force can halt the continuing collapse. General relativity predicts its collapse to a massive, infinitesimal point, a "singularity." Physicists shun singularities, and quantum theory would replace the singularity with an extremely compact, but finite-sized, mass in some not-yet-understood way.

Within a distance from this compact mass, which could be many kilometers, inside the so-called "event horizon," the gravitational attraction is so great that not even light can escape. This collapsed star thus emits no light. It's therefore black. Anything venturing inside the horizon can never get out. It's a black hole.

Stephen Hawking showed that quantum mechanics enters the black-hole picture not only at the singularity but also at the horizon. Quantum effects should cause the black-hole horizon to emit what is now called

"Hawking radiation." Emitting energy, any black hole that cannot pull in mass from its surroundings would eventually radiate away, or "evaporate" and disappear.

Though the evaporation time scale for large black holes could be longer than the age of the universe, black hole evaporation has raised a paradox. Quantum theory insists that the total amount of "information" is always preserved. (Can the concept of "information" be independent of the concept of "observation"?) But if the Hawking radiation were random thermal radiation, as initially thought, all the information that had been contained in objects falling into a black hole would be lost when the black hole evaporated.

We're using a far-fetched notion of "information" here. If, for example, you throw your diary into a fire, someone can, in principle, recover its information by analyzing the light, smoke, and ashes. The apparent quantum-theory-violating loss of information in black-hole evaporation led Hawking to speculate that the information might, when the hole evaporated, be channeled to a parallel universe.

Hawking recently decided that a black hole's radiation is *not* random, that the radiation actually carries off the information contained in objects that have fallen into the hole—the same way smoke carries off information from your burning diary. No need for parallel universes to take up black-hole information. Nevertheless, some cosmologists, for other quantum reasons, suggest that ours is likely not the only universe, and even talk of, albeit weak, observational evidence for that.

Black holes got attention in the popular press in 2009 when a group petitioned the United Nations to prevent the startup of the $5.5 billion Linear Hadron Collider (LHC) near Geneva, Switzerland. The fear was that the machine, colliding protons at never-yet-achieved energies (14 TeV) would create a black hole, which would then gobble up the planet. The possibility of creating *tiny* black holes had indeed been theoretically postulated, but they would presumably quickly and harmlessly evaporate. A committee of physicists was actually appointed to study and respond to the concern. Their convincing argument for no danger was that our planet has long been bombarded by cosmic rays with energies of the LHC and much higher. And we are still here. The LHC has now turned on. No black holes so far.

Dark Energy

Modern cosmology is based on Einstein's theory of *general* relativity. It is "general" in the sense that it extends his earlier special relativity to include accelerated motion and gravity with the realization that the two are equivalent. For example, should the elevator cable break, your downward acceleration would cancel your experience of gravity.

Though mathematically complex, general relativity presents a conceptually beautiful, straightforward theory. However, the form Einstein first wrote down in 1916 seemed to have a serious problem. It said the universe could not be stable. The mutual gravitational attraction of the galaxies would cause them to collapse in on themselves. Einstein patched up his theory by adding the "cosmological constant," a repulsive force to counter the gravitational attraction.

In 1929 astronomer Edwin Hubble announced that the universe was *not* stable. It was, in fact, expanding. The more distant a galaxy was from its neighbors, the faster it moved away from them. If so, some time in the past everything was clumped together. That gave rise to the idea of the universe starting with a big explosion, the Big Bang. The galaxies are therefore still flying apart. That could explain why galaxies did not fall in on each other. No repulsive force, no cosmological constant, was needed.

An explosion is not quite the right picture. General relativity has space *itself* expanding, not galaxies flying apart in a fixed space. Specks of paper pasted on an inflating balloon, and thus moving apart faster the more distant they are from each other, is a good analogy.

When Einstein realized that the universe was indeed *not* stable, he threw out his cosmological constant, calling it the "greatest blunder of my career." If he had only believed his original, more beautiful, theory he could have predicted an expanding (or contracting) universe more than a decade before its observational discovery.

The gravitational attraction of the galaxies for each other should slow the expansion, just as gravity slows an upwardly thrown stone as it rises. The stone rises to some height and then falls back down. Similarly, one might expect the galaxies to slow down, reach some maximum separation, and eventually fall back together in the Big Crunch.

If you throw a stone up *fast* enough, it will continue out in space forever. However, still pulled back by Earth's gravitational attraction, it will continually slow. By the same token, if the Big Bang were violent enough, the universe would expand forever, albeit at a slower and slower rate. By determining the rate at which an upwardly thrown stone is slowing, you can tell whether it will fall back down or continue out forever. By finding the rate at which the expansion of the universe slows, we can tell whether or not to expect the Big Crunch.

Actually, it has been recognized for a couple of decades that the galaxies do not constitute all the mass of the universe, not even the largest part. The motions of stars within galaxies, and other evidence, tell us that there is a kind of matter out there in addition to the stuff that the stars, the planets, and we are made of. It has gravitational attraction but does not emit, absorb, or reflect light. We thus cannot see it. It's "dark matter." No one knows what it is, but people have built detectors to search for the likely suspects. It's the sum of the normal matter and the dark matter that would be expected to slow the expansion to determine the eventual fate of our universe.

(On a recent PBS *Nova* program, an astronomer said he could not think of a more fundamental question for humankind than: "What is the eventual fate of our universe?" Perhaps this is indeed a pressing question. But it recalls a story: In a public lecture, an astronomer concluded: "Therefore, in about five billion years the sun will expand as a red giant and incinerate the inner planets, including Earth." "Oh, no!" moaned a man in the rear. "But, sir, it won't happen for another *five billion years*," reassured the astronomer. The man's relieved response was, "Oh, thank God! I thought you said five *million* years.")

In the past decade, astronomers set out to determine the fate of the universe by measuring how fast certain distant exploding stars, supernovas, are receding. These particular explosions have a characteristic intrinsic brightness, and therefore astronomers can tell how far away they are by how bright they appear. The farther away they are, the longer ago the light we now receive must have left them. Putting all this together, they could determine how fast the universe was expanding at different times in the past, and therefore determine the rate of slowing.

Surprise! The expansion of the universe is *not* slowing. It's *accelerating*. Not only is the mutual gravitational attraction of the galaxies canceled, but there is a repelling force in space that *exceeds* the gravitational attraction. With that force must come an energy.

Since mass and energy are equivalent ($E = mc^2$), this mysterious repulsive energy has a mass distributed in space. In fact, *most* of the universe is made up of this mysterious "dark energy." The universe appears to be about seventy percent dark energy and twenty-five percent dark matter. The kind of stuff the stars, the planets, and we are made of appears to be a mere five percent of the universe.

Though no one knows what the dark energy is, in a formal sense it brings Einstein's cosmological constant, his "biggest blunder," back into the equations of general relativity. Theoretical guesses have an uncanny way of ending up right.

Is it conceivable that the mysterious dark energy involves a connection between the large-scale universe and consciousness that Martin Rees's comment at the start of this chapter might imply? Hardly. But let's quote the quantum theorist Freeman Dyson, writing even before the idea of dark energy arose:

> It would not be surprising if it should turn out that the origin and destiny of the energy in the universe cannot be completely understood in isolation from the phenomena of life and consciousness. . . . It is conceivable . . . that life may have a larger role to play than we have imagined. Life may have succeeded against all odds in molding the universe to its purposes. And the design of the inanimate universe may not be as detached from the potentialities of life and intelligence as scientists of the twentieth century have tended to suppose.

The Big Bang

Astronomers determine the speed with which a galaxy recedes from us by the "redshift" of its light. This frequency lowering is similar to a "Doppler shift," the lowered pitch of the siren of an ambulance that has just passed us. It's actually the expansion of space stretching the light's wavelength.

Astronomers correlate an object's redshift with its distance from us by studying the redshifts of objects whose absolute brightness, and therefore distance from us, is known. They find that the most distant objects we can see, galaxies moving away from us at close to the speed of light, emitted the light we now receive some thirteen billion years ago. Those galaxies were probably about one billion years old when that light was emitted. This suggests that the Big Bang occurred about fourteen billion years ago.

By the time the universe was 400,000 years old, it had cooled enough to allow the light-scattering electrons and protons to combine into neutral atoms, and for the first time the universe became transparent to the radiation created in the initial fireball. Radiation and matter in the young universe thus became independent of each other. At this point the radiation, initially at very high frequency, was largely in the ultraviolet and visible region of the spectrum. But since that time, space has expanded more than a thousand-fold. The wavelength of that first light has now been stretched by that factor to become the three degree Kelvin "cosmic microwave background" now shining down on us from all directions. This microwave radiation, accidentally discovered in 1965 by physicists at AT&T's Bell Laboratories who were studying communications satellites, is the strongest evidence for the Big Bang. Its fine details strikingly confirm properties calculated for the Big Bang.

Theories of "inflation" speculate about the *immediate* aftermath of the Big Bang to explain the striking uniformity of the universe on the largest scales, as evidenced in the distribution of the galaxies and the microwave background radiation. According to these ideas, space almost instantly expanded, or "inflated." Parts of it moved away from each other at a rate much faster than the speed of light. This does not violate special relativity's speed limit being that of light. During inflation objects were not moving *in* space faster than light. Objects were getting farther apart because space itself was expanding. Starting from something vastly smaller than an atom, the entire universe we observe today presumably inflated almost instantaneously to the size of a large grapefruit.

An interjection: We're obviously talking of an epoch as far from conscious observers as we can imagine. One might think that the experts studying this physics would hardly involve themselves with considerations of consciousness. Not necessarily so. In one of the most important books on the subject, *Particle Physics and Inflationary Cosmology* (not easy reading), Stanford University physics professor Andrei Linde writes:

> Will it not turn out, with the further development of science, that the study of the universe and the study of consciousness will be inseparably linked, and that ultimate progress in the one will be impossible without progress in the other? . . . will the next important step be the development of a unified approach to our entire world, including the world of consciousness?

In a recent video interview (http://www.closertotruth.com/video-profile/Why-Explore-Consciousness-and-Cosmos-Andrei-Linde-/874) Linde tells that his editor suggested he remove the reference to consciousness in his book because he "might lose the respect of his friends." Linde told her that if he removed it, "I would lose my own self-respect."

After the ridiculously short period of inflation, physics seems able to account for what happened in some detail. By the time the universe was one second old, quarks combined to form protons and neutrons. A few minutes later the protons and neutrons came together to form the nuclei of the lightest atoms: hydrogen, deuterium (heavy hydrogen, one proton and one neutron), helium, and a bit of lithium. The relative abundance of hydrogen and helium in the oldest stars and gas clouds agrees with what we would expect from this creation process.

But during that split second before our "familiar" quarks and electrons came into existence, the Big Bang had to be finely tuned to produce a universe in which we could live. *Quite* finely tuned! Theories vary. According to one, if the initial conditions of the universe were chosen randomly, there would only be one chance in 10^{120} (that's one with 120 zeros after it) that the universe would allow life. Cosmologist and consciousness theorist Roger Penrose has it vastly more unlikely: The *exponent* he suggests is 10^{123}. (It's hard to comprehend the meaning of a number that big.) By any such estimate, the chance that a livable universe like ours would be created is far less than the chance of randomly picking a *particular* single atom out of all the atoms in the universe.

Can you accept odds like that as a coincidence? It might seem more likely that something in yet-unknown physics determines that the universe *had* to start the way it did. Such new physics would likely include a quantum theory of gravity. It might well be the long-sought "theory of everything," the ToE, uniting Nature's four fundamental forces into a single theory. All physical phenomena should then be explainable, in principle at least.

We know what the ToE will look like. It will be a set of equations. After all, that's what the searchers are seeking. Could a set of equations resolve the quantum enigma? Recall that physics' encounter with consciousness is seen directly in the *theory-neutral* quantum experiment. It arises logically *prior* to the quantum *theory*, from assumptions including free will. An interpretation of quantum theory, or even its deduction from a more general mathematical presentation, could therefore not resolve the quantum enigma without somehow involving our conscious decision process.

With a perhaps similar perspective on whether a ToE would explain what we see, Stephen Hawking poses a question:

> Even if there is only one possible unified theory, it is just a set of rules and equations. What is it that breathes fire into the equations and makes a universe for them to describe? The usual approach of science of constructing a mathematical model cannot answer the questions of why there should be a universe for the model to describe. Why does the universe go to all the bother of existing?

Some suggest that an eventual ToE will predict all we see, even though it will not "explain" it. We must therefore seek the ToE as an ultimate goal and be satisfied with it if we find it. That's all we can expect of science. That's also the attitude the two of us accept—most days. But not always.

Critics of this attitude speak of an anthropic principle. We start with the more easily accepted version, but warn of wilder ideas as we bring our book to a close.

The Anthropic Principle

Only the very lightest nuclei were created in the Big Bang. The heavier elements, carbon, oxygen, iron, and all the rest, were created inside stars, which formed much later. These elements beyond hydrogen and helium are released into space when a massive star, exhausting its nuclear fuel, violently collapses, and then explodes as a supernova. Later-generation stars and their planets, including our solar system, gather up this debris. We are the remnants of exploded stars. We're stardust.

In addition to the extreme fine-tuning of the Big Bang that we just mentioned, another bit of luck seems involved in our stellar creation. Early calculations had shown that the making of heavy elements in stars could not get even as far as the carbon nucleus (six protons and six neutrons). Cosmologist Fred Hoyle reasoned that, since carbon was indeed here, there *had* to be a way of making it. He realized that a then-unexpected quantum state of the carbon nucleus at a certain very precise energy could allow the stellar production of the elements to continue to carbon, nitrogen, oxygen, and beyond. Hoyle suggested that the totally unexpected nuclear state be looked for. It was *found*.

There are other coincidences: if the strengths of the electromagnetic and gravitational forces were even *slightly* different from what they are, or if the strength of the weak nuclear force were *slightly* larger, or *slightly* smaller, the universe would not have been hospitable to life. No known physics compels these things to work out just right.

Other coincidences have been noted beyond those we mention. Do things working out so perfectly, but so improbably, require explanation? Not necessarily. If it didn't just *happen* to work out just this way, we wouldn't be here to ask that question. Is that explanation enough? Such backward reasoning, based on the fact that we and our world exist, is called the "anthropic principle."

The anthropic principle can imply that our universe welcomes life just by chance. On the other hand, some theorize that a large number of universes, even an infinite number, came into existence, each with its own random initial conditions, even with its own laws of physics. Some theories have a grand "multiverse" constantly spawning new universes. The vast majority of these universes likely have a physics that is not life-friendly. Does our improbable existence in a rare, hospitable one therefore need explanation?

Here's an analogy: Consider how improbable *you* are. Consider the unlikelihood of you, someone with your unique DNA, being conceived. Millions of your possible siblings were *not* conceived. Now go back a few generations. With those odds, you're essentially impossible. Does your being here require explanation?

With analogies like this, some urge science to shun the anthropic principle. The anthropic principle, they claim, explains nothing. It should therefore be rejected as "needless clutter in the conceptual repertoire of science." Arguably, it can have a negative influence by dampening the drive for deeper searches. But anthropic reasoning can sometimes be fruitful. Consider Hoyle's energy-level prediction for carbon.

Objectors to the anthropic principle, what we now can call the "weak anthropic principle," might be even more averse to the "strong anthropic principle." According to this view, the universe is tailor-made for us. "Tailor-made" implies a tailor, presumably God. That may be something to contemplate. But it's hardly an argument for Intelligent Design, as is occasionally suggested. Whoever "breathes fire into the equations" would presumably be omnipotent enough to do it properly at the very beginning and not need to tinker with every step of evolution.

A different version of the strong anthropic principle is implied in our quote at the start of this chapter: *We* created the universe. Quantum theory has observation creating the properties of microscopic objects, and we generally accept that quantum theory applies universally. If so, may wider reality also be created by our observation? Going *all* the way, *this* version of the strong anthropic principle asserts the universe is hospitable to us because we could not create a universe in which we could not exist. While the weak anthropic principle involves a backward-in-time reasoning, this strong anthropic principle involves a form of backward-in-time *action*.

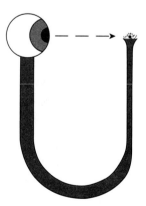

Quantum cosmologist John Wheeler back in the 1970s drew an eye looking at evidence of the Big Bang and asked: "Does looking back 'now' give reality to what happened 'then'?" His provocative sketch has not lost impact. At the recent conference I (Fred) attended, honoring Wheeler on his 90th birthday, a keynote speaker introduced his talk with Wheeler's sketch.

Figure 18.1 Does looking back "now" give reality to what happened "then"?

The anthropic implications of Wheeler's sketch must have been a lot even for Wheeler to buy. After his "looking back" question, he immediately added the comment: "The eye could as well be a piece of mica. It need not be part of an intelligent being." Of course, that piece of mica, supposedly bringing reality to the Big Bang, was presumably created *after* the Big Bang. (For a physicist, it can be less disturbing for a piece of mica to create the Big Bang than conscious observation.)

This strong anthropic principle is probably too much for anyone to believe, or even comprehend. If our observation creates *everything*, including ourselves, we are dealing with a concept that is logically self-referential and therefore mind-boggling.

Accepting the boggle, we might ask: Though we could only create a universe in which we could exist, is the one we did create the only one we *could* have created? With a different choice of observation, or a different postulate, would the universe be different? It has been wildly speculated that merely postulating a theory that is not in conflict with *any* previous observation actually *creates* a new reality.

For example, Hendrick Casimir, motivated by the discovery of the positron after its seemingly unlikely prediction, mused: "Sometimes it almost appears that the theories are not a description of a nearly inaccessible reality, but that so-called reality is a result of the theory." Casimir may also have been motivated by his own prediction, later confirmed, that the quantum mechanical vacuum energy in space would cause two macroscopic objects to attract each other.

Just for fun: If there were anything to Casimir's conjecture, might Einstein's original suggestion of a cosmological constant have *caused* the acceleration of the universe? (Such a speculation can't be *proven* wrong. It's therefore not a *scientific* speculation.) Though taking an idea like this literally is surely ridiculous, the quantum enigma can motivate outrageous speculation.

John Bell tells us that the new way of seeing things will likely astonish us. It is hard to imagine something truly astonishing that we don't initially rule out as preposterous. Bold speculation may be in order, but so is modesty and caution. A speculation is nothing but a guess until it makes testable and confirmed predictions.

Parting Thoughts

We have presented the quantum enigma arising from the brute facts displayed in undisputed quantum experiments. We do not presume to resolve the quantum enigma. The questions the enigma raises are more profound than any resolution we could seriously propose.

Quantum theory works perfectly; no prediction of the theory has ever been shown in error. It is the theory basic to all physics, and thus to all science. One-third of our economy depends on products developed with it. For all *practical* purposes, we can be completely satisfied with it. But if you take quantum theory seriously *beyond* practical purposes, it has baffling implications.

Quantum theory tells us that physics' encounter with consciousness, which is *demonstrated* for the small, applies, in principle, to everything. And this "everything" can include the entire universe. Copernicus dethroned humanity from the cosmic center. Does quantum theory suggest that, in some mysterious sense, we *are* a cosmic center?

The encounter of physics with consciousness has troubled physicists since the inception of quantum theory more than eight decades ago. Many, no doubt most, physicists dismiss the creation of reality by observation as having little significance beyond the limited domain of the physics of microscopic entities. Others argue that Nature is telling us something, and we should listen. Our own feelings accord with Schrödinger's:

> The urge to find a way out of this impasse ought not to be dampened by the fear of incurring the wise rationalists' mockery.

When experts disagree, you may choose your expert. Since the quantum enigma arises in the simplest quantum experiment, its essence can be fully comprehended with little technical background. Non-experts can therefore come to their *own* conclusions. We hope yours, like ours, are tentative.

> There are more things in heaven and earth, Horatio,
> than are dreamt of in your philosophy.
>
> —Shakespeare, *Hamlet*

Suggested Reading

Baggott, J., *The Quantum Story: A History in Forty Moments*. New York: Oxford University Press, 2011.

> An excellent treatment of the turbulent history of quantum mechanics by an author who is not only technically expert in the theory, but one who appreciates its profound implications..

Bell, J. S., and Aspect, A., *Speakable and Unspeakable in Quantum Mechanics*, 2nd ed. London: Cambridge University Press, 2004.

> The collected papers of John Bell. Most are quite technical. But even these have parts that, with wit, display the insights of the leading quantum theorist of the last half of the twentieth century.

Blackmore, S., *Consciousness: An Introduction*. New York: Oxford University Press, 2004.

> A wide-ranging overview of the modern literature of consciousness from the neural correlates of consciousness, to experimental and theoretical psychology, to paraphenomena. Some mentions of quantum mechanics are included.

Cline, B. L., *Men Who Made a New Physics*. Chicago: University of Chicago Press, 1987.

> This light, well-written history of the early development of quantum mechanics emphasizes the biographical and includes many amusing anecdotes. Since it was originally written in the 1960s, it avoids any significant discussion of the quantum connection with consciousness. (One of the "men" is Marie Curie.)

Davies, P. C. W., and J. R. Brown, *The Ghost in the Atom*. Cambridge: Cambridge University Press, 1993.

> The first forty pages give a compact, understandable description of "The Strange World of the Quantum." This is followed by a series of BBC Radio 3 interviews with leading quantum physicists. Their extemporaneous comments are not always readily understandable, but they clearly give the flavor of the mystery they see.

d'Espagnat, B., *On Physics and Philosophy*. Princeton, N.J.: Princeton University Press, 2006.

> An extended, authoritative treatment going deeply into the issues of reality and consciousness raised by quantum mechanics. No mathematical jargon, but not easy reading.

Elitzuir, A., S. Dolev, and N. Kolenda, eds., *Quo Vadis Quantum Mechanics?* Berlin: Springer, 2005.

> A collection of articles, and transcripts of informal discussions, by leading researchers with an emphasis on the paradoxical aspects of quantum mechanics. Some of the papers are highly technical. But aspects of several are quite accessible and indicate how physics has encountered what seems a boundary of the discipline.

Gilder, L., *The Age of Entanglement: When Quantum Physics Was Reborn*. New York: Vintage Press, 2009.

> Imagined conversations among the founders of quantum theory, all based on well-documented sources, and real conversations with recent workers. Engaging and easy reading.

Griffiths, D. J., *Introduction to Quantum Mechanics*. Englewood Cliffs, N.J.: Prentice Hall, 1995.

> A serious text for a senior-level quantum physics course. The first few pages, however, present interpretation options without mathematics. The EPR paradox, Bell's theorem, and Schrödinger's cat are treated in an "Afterword." (The book's front cover pictures a live cat; it's dead on the back cover.)

Hawking, S., and L. Mlodinow, *A Briefer History of Time*. New York: Random House, 2005.

> A brief, easy-to-read, but authoritative, presentation of cosmology, much of it from a quantum mechanical point of view. Metaphysics and God get substantial mention.

Holbrow, C. H., J. N. Lloyd, and J. C. Amato, *Modern Introductory Physics*. New York: Springer, 1999.

> An excellent introductory physics text with a truly modern perspective, including the topics of relativity and quantum mechanics.

Kumar, M., *Quantum: Einstein, Bohr, and the Great Debate about the Nature of Reality*. W. W. Norton & Company, 2010.

> An authoritative and interestingly written treatment of Einstein-Bohr "Debate." Bell's theorem and the continuing debate are more briefly presented.

Miller, K. R., *Finding Darwin's God: A Scientist's Search for Common Ground between God and Evolution*. New York: HarperCollins, 1999.

> A convincing refutation of Intelligent Design that also argues that arrogant claims of some modern scientists that science has disproven the existence of God has promoted antipathy to evolution, both Darwinian and cosmological. Quantum mechanics plays a prominent role in Miller's treatment.

Park, R. L., *Voodoo Science: The Road from Foolishness to Fraud*. New York: Oxford University Press, 2000.

> A brief, cleverly written exposure of a wide range of purveyors of pseudoscience who exploit the respect people have for science by claiming that science gives credence to their particular nonsense.

Schrödinger, E., *What Is Life?* and *Mind and Matter*. London: Cambridge University Press, 1967.

> An older but very influential collection of essays by a founder of quantum theory, including one titled "The Physical Basis of Consciousness."

Index

absolute square, of wavefunction, 79
absolute zero, 121
absurdity, of quantum theory, 143–45,
    147–48
acceleration, 28–230
AI. *See* artificial intelligence
air resistance, 24–25
alchemy, 34
Alfonso X (king of Castile), 23
algae, 198
Alice and Bob thought experiments, 159–60,
    164–70, 165*f*, 179–85, 187–90,
    200–201
Allen, Woody, 34
alpha particles, 66–67, 67*f*, 208
    waviness and, 80–81, 81*f*
American Physical Society, 197
American Revolution, 43
angular momentum, 68
anthropic principle, 265–68
Aquinas, Thomas, 22
arbitrage modeling, 204
argon atoms, 139*f*
Aristotle, 21–23, 27–28, 32, 34
    Galileo and, 24–26, 90
    Metaphysics, 102, 193
    philosophy of, 25–26
    Physics, 102, 193
artificial intelligence (AI), 235, 248
Aspect, Alain, 186, 192
The Astonishing Hypothesis (Crick), 231
astronomy, 119
    Medieval, 22–24
    radio, 184
atom bombs, 173
atomic nature, of matter, 61, 65
atomic-scale objects, 130–31
atoms, 4, 61–62
    of argon, 139*f*
    atoms fired one at a time through a
        movable two-slit barrier, 157*f*

bouncing a photon off an atom, 130*f*
    of carbon, 80
    distribution through single narrow slit, 89*f*
    electric charges of, 43, 46
    excited state of, 163
    ground state of, 163
    of hydrogen, 48, 66, 68, 70, 79–80,
        79*f*, 120–21
    impurity atoms, 119
    in interference patterns, 88*f*, 89*f*
    internal structure of, 65–68
    north and south poles of, 99, 121
    rice pudding model of, 66–67, 66*f*
    Rutherford's atomic model, 67*f*
    scanning tunneling microscopes
        for, 92, 139
    unobserved and observed, 130–31
    wavefunction of, 77–79, 83, 89, 91*f*
Australia National University, 229
awareness neuron, 231

Bayesian probability, 255
behaviorist psychology, 221, 228–29
Bell, John Stewart, 175*f*, 202, 217, 226, 268
    background of, 174–75
    motivation of, 176, 206, 213
    spooky actions and, 174–76
    Weinberg and, 101
Bell's inequality
    derivation of, 178–84, 179*f*, 180*f*,
        181*f*, 182*f*, 183*f*
    physical reality and, 187–88
    violation of, 184, 186–87, 190
Bell's theorem, 171, 201, 211
    experimental tests of, 184–87, 240
    spooky actions and, 177–78
Bem, Daryl, 254
Bennett, Bill, 3
benzene, 118
benzol, 132
Berkeley, George, 227, 253